the light through the trees

REFLECTIONS ON LAND AND FARMING

LUANNE ARMSTRONG

01 02 03 04 05 06 17 16 15 14 13 12

Caitlin Press Inc.
8100 Alderwood Road,
Halfmoon Bay, BC von 1y1
www.caitlin-press.com

Text and cover design by Kathleen Fraser.
Edited by Jane Hamilton Silcott.
Cover photo by Dorothy Woodend.
Back cover and text photos by Avril Woodend.
Printed in Canada.

Caitlin Press Inc. acknowledges financial support from the Government
of Canada through the Canada Book Fund and the Canada Council
for the Arts, and from the Province of British Columbia through the
British Columbia Arts Council and the Book Publisher's Tax Credit.

 Canada Council   Conseil des Arts
for the Arts   du Canada

 BRITISH COLUMBIA
ARTS COUNCIL
An agency of the Province of British Columbia

Library and Archives Canada Cataloguing in Publication

Armstrong, Luanne, 1949–
        The light through the trees : reflections on land and
farming / Luanne Armstrong.

ISBN 978-1-894759-95-3

        1. Farm life—British Columbia. 2. Country life—British
Columbia. 3. Human ecology. I. Title.

S522.C3A75 2012   630.9711   C2012-903334-0

THE LIGHT THROUGH THE TREES
REFLECTIONS ON LAND AND FARMING

Luanne Armstrong

Dedicated with much love to my brother Bill and sister, Robin, with whom I share the farm.

Contents

Fresh Tracks

THE DOG KNOWS. HE STANDS beside my desk, his tail wags very slightly, his fierce collie stare fixed on my face.

Yes, yes, all right. I have been in front of the computer for several hours, teaching a class of writers who live all over the world. I live in the Purcell Mountains, beside the broad, dark expanse of Kootenay Lake. I have lived here on the same piece of land for most of my life except for time out to travel, teach and go to school.

Last night it snowed. Time to go outside, read the new snow, see who has been here, who passed through, who came to visit in the night.

We set off—my dog, my brother's dog and I—down the hill and across the field below my house. The coyotes were here this morning hunting mice in the tall grass. There are deer tracks, raven tracks, and an unknown cat track in the snow. Probably bobcat.

The two black dogs run and run. In *Inside of a Dog*, Alexandra Horowitz writes that dogs see time differently than humans do.[1] The dogs don't see tracks, they see smells. They know who was here today, and last week. The only time I can somewhat "see" in the way the dogs do is when I look at the layers of tracks in the snow laid down over the past couple of weeks. Today, the fresh snow has obscured some tracks and revealed others.

And then on down the hill toward the beach where bear tracks cross the path. Late in the year for a bear to be out.

On the yellow sand, edged with ice, more deer tracks, evidence that several deer have been here. From the tracks they left, they might have been dancing. Playing? I don't know.

I stand to admire the glassy sheen of pale water, black rocks crusted with ice, the mountains stippled pale blue and bright white. Two black and white ducks at the point of a thin V on the surface. Few people come to the lake in winter. They miss this beauty. On calm days in winter, the water surface looks glazed, like metal. Just past the beach, a large bald eagle lights in a Douglas fir. The top bends under its weight. The eagles come in winter to feed on the black coots. We trudge up from the beach, up the hill, over the fence and across the pasture and so, home again to tea and darkness coming and some more writing done.

Writing and walking go well together. Walking is always a story, and though I have taken this same walk almost every day for sixty years, each walk tells me a different story. Seasons change, colours change. I meet the other inhabitants of the place or read their stories in tracks, in sounds, in flashes of colour.

When I am writing, and stuck for a new direction, an idea, new words, walking usually brings an idea or some language to get me through to a new place. As Rebecca Solnit says in *Wanderlust: A History of Walking*, "Walking shares with making and working that crucial element of engagement of the body and the mind with the world, of knowing the world through the body and the body through the world."[2]

Walking here unites all the disparate elements of my world, mind and body, human and animal, the past and the present, inside and out. Every walk is a narrative about language and silence, about presence and about forgetting. When I walk, I travel through a place I love and am trying to understand.

Writing about love is hard. A love story tends to always veer into romance, or sentiment, or lyrical grandiloquence. And yet a love story is impossible to avoid. It wants to be told. It trumpets its own eloquence.

Is love even the right word? I write about feeling connected, about the many ways I see this place. And about the many eyes here that also see me? And I also write about what I don't know. What do these non-human others know? What do they see? And how can we see each other past this not-knowing?

I wonder if this is even a relationship, my feelings about this place and its inhabitants. Or is it all a one-way emotion from me, the human stalker, wandering around wanting to be loved? And just what am I doing here, living in this particular place? How do I understand it?

I have asked myself these questions at many odd and various moments. While I bend over the garden, planting. In the spring, breathing on tulips, or listening to a lone frog, both of us awake on a March night. Or in August, listening to the Northern harrier crying over the burned-to-golden summer field as I sit peeling peaches in the hot slanting sun. Watching my farmer self, at harvest, harried by wild turkeys in the grapes, deer in the apple trees, voles in the garden, and the half-grown bear in the pear tree. In the winter mornings, a raven comes by as I throw hay to the cows, standing ankle deep in yellow mud and manure.

At night, the dogs and the coyotes yell challenges or greetings or some other message. I listen to exchanges from which I am excluded. What do I know about all this?

Around and around we go, a palimpsest of footprints telling an infinite number of stories, over my lifetime, over so many lifetimes.

If I put my ear to the ground, if I lie down, can I hear the past banging its way under the grass and tree roots? Can I hear the banging of all those other feet coming by? Can I hold eternity by the hand like a child with almost no sense of myself, listening, at last, inside this place? If I weren't walking here, would I be walking somewhere else, my head in the sky and my feet shuffling in grass, in leaves? Wondering how to grasp the intricate complexity of a field, a patch of moss, a flower opening. What do I really know?

This fall, I missed the swallows leaving and felt an acute sense of loss. I was interrupted by inattention, by being busy. No excuses. And the ospreys left as well, without saying goodbye. No, it was me who didn't say goodbye, stomping around picking apples and preparing for my own winter. Another year going around—all year we chase each other, the seasons and I, round and round. I am lost inside and lost outside, occasionally glad to be so lost, and yet often too anxious, driven by time and small worries.

The elements bind us, me, the place, its many inhabitants, together—fire inside my belly, fire in the bellies of the sleepy animals, fire in the woodstove, and then Kootenay Lake glinting in its cleft home between mountains, my feet banging, banging on the earth while I listen to echoes.

In winter, late at night, I curl under the weight of many covers, listening to snow hissing at the windows, wind shivering the tree branches together in an odd syncopation, in my bed house, my bed-balloon, my bed cocoon, tethered to the night sky, swinging and whirling in the wind, travelling all night but never lost, never not at home.

Walking here is also walking in time, over layers of tracks laid down by parents and grandparents. Understanding how to fully be in a place and taking care of it means also seeing how it is taking care of me.

Inside these reflections is a writer walking in slow circles, through one place, over and over, studying her own tracks to find not only how to live somewhere but also to love the place where she lives, to care for it, and to try to understand, as much as is humanly possible, the place and its inhabitants. These essays attempt to illustrate, open, and perhaps begin, however tentatively, to illuminate those gnarled and ancient questions that I share with anyone who has a deep relationship with place: what am I doing, how do I live well here, how do I behave within this network of relationships, both human and not?

I have called myself a farmer for most of my life, but I am also a writer. Which means it's hard to be in a relationship with land, animals, a particular place and the people of the place without wanting to tell stories about it and wondering what it all means.

But I'm not the only one asking these kinds of questions. I write these essays at a time when the tangled and contradictory arguments in the media and in environmental circles about the future of the world—about global warming, about energy shortages, about food panics—continue with no resolution.

Asking questions about a relationship with a place, and telling stories about that relationship is simple but necessary in a world where the assumptions upon which humankind has prided itself—the idea of progress, the idea of human superiority over the non-human, the idea of private property—are demanding to be questioned.

Changing ways of thinking and talking about place and our relationships to them will only happen through a sustained conversation and communal cultural examination of assumptions and priorities. These essays are an ongoing part of such a conversation.

And although many writers live in places they love, few are farmers and even fewer write about farming. The books about farming and gardening that I have read usually follow two standard formats. Either they are practical "how to" books, or they are written by non-farmers about farming. Thus farm writing, like nature writing, has a tendency to be both sentimental and romantic. But I was a farmer before I was a writer. So here are stories of a small farm, of the stubborn, self-sufficient, hard-working and proud way of life that I was born into, and which I continue. When I was a child, I worked on the farm because my father told me that I had to work or we would starve. Now I do it because nothing else I have done is more satisfying. I like growing food and sharing it, I like animals, I like walking, and I find in the place where I live an endlessly engrossing and unfolding story.

Once I was talking to a dear friend about my childhood and the farm, and he said, "The land was your mother and father." I think he was right, but as well as mother and father, I think the land is also both my teacher and my home, the place I go to learn, the place that mystifies me with its depth and beauty, and never lets me go.

Turnings

Wolves surprised us one afternoon on the mountain, howling on their way through from somewhere to somewhere else.

A cougar tracked through the pasture in the last snow. Then the other day my brother came in with a hunk of grey hair. Something had broken the bottom two strands of barbed wire in the north fence and left a chunk of hair behind. We looked at it mystified—not deer, not moose, not anything we knew. The next morning, he came back, gleeful, having talked to a biologist. "A grizzly," he announced. Not everyone would be thrilled at the sign of such large predators around, but we are glad at this sign of a functioning and intact ecosystem.

Partly it is because there are fewer hunters and fewer people here in the winter. Driving along the lake road means driving by many huge and shuttered houses. Even the few people who still live here year-round tend to go away for big chunks of the winter. Even so, many of the houses still sport large yard lights.

From my house at night, in a landscape that was once pristinely dark, I can now see three sets of lights. They burn all night and every night, and when I look out, they irritate me like an itch I can't scratch.

But mostly the farm is quiet. The pigs are gone to people's freezers. The empty pigpen is sad. The garden is asleep under the snow. The greenhouse is shut down. A few birds eat dried Saskatoon berries and rose hips. Flickers occasionally come to drill the logs of the house for dinner. An eagle goes by on its way to harass the coots but the ravens don't chase it as they usually do.

Walking is an experience of black and white and grey: grey water sloshing restlessly under the wind, black rocks, black trees, white snow layered on every surface.

It's the social time of the year for me, and conversely, also the quiet time: time to write, to think, to walk. Christmas, Yule, Solstice, the turning of the year, the time when there is still abundance left from harvest—when the canning jars, the dried fruits, the boxes of nuts and onions and garlic are still here. The hungry time of the year is still to come. No wonder this time of year is a festival of life and Easter is a festival of death. Traditionally, Easter would have come at a time when the cupboard was empty and the garden not grown.

But it's Christmas; all is well at Kootenay Lake. I write and listen to the radio and read and study. And I watch, listen and wait. And while I do that, I plan the next garden.

My Father's Gift

ABOUT A YEAR BEFORE MY father died, on one of my trips home to our farm, I got up one morning and my father offered me two pieces of fruit for breakfast, an orange and a grapefruit. I had heard him come in—he still got up at 5 a.m. and went out to feed the animals. I said, "Where did you get it?"

"Your brother brought it out for the pigs, but it's only a bit rotten, you cut that part off, it's still good."

"I don't eat pig food."

"It's got a hide on it," he said. "What's the matter with you?"

Words between my father and me were always strung tight, dangerous as barbed wire. The land my father and I once shared, and where I now live again, is layered and slick with these words, some of them like varnish, some like glue. We have written our family, our history, on this land, in buildings, fences and stories. After our father died, my brother and I took over the farm, gardening, feeding animals and living much as our father lived. But with some major differences. We love this place. Our father never did.

My father fought to subdue the land. He bulldozed and dug and drained and cut down trees and blew up boulders and sprayed weeds and insects. He made things grow by the strength of his unrelenting will, and we, his children, ate his food and sat at his table and worked his land with him.

As I became what I am now—a writer, a feminist, an academic—my every gesture, my every act—reading, writing, speaking, leaving—enraged him. He had rarely given me anything as a gift, and that morning, he was proud of this free food he was holding out to me. What could I do with such a gift?

Much later, I was sitting in a class on women's autobiography. We were talking about gifts and the meaning of gifts. When I told them about my father's gift, a young woman across the circle made a face. "He's demented," she said.

But he wasn't demented. My father, by his terms, was eminently reasonable and his gift made sense. As his daughter, I understood very well that this gift was another part of the confused and confusing messages that had always passed between us, part of the hidden war that had been going on ever since I became a girl. I understood his independence, his disdain of "storebought," his glee at the idea of free food. Along with the reasons for his cheapness, I understood that what he was offering was a true gift, but a gift, like everything my father had ever given me, with razored edges.

For most of their lives, my parents survived as self-sufficient small farmers, a life and a culture that was once common in Canada and now has almost disappeared. Our farm was then about twenty acres of rocks, swamp and hayfield beside Kootenay Lake, plus another sixty acres of mountainside above the highway. When I was young the farm produced almost all our food. We had a half acre of garden and an apple, cherry, plum and pear orchard that the original homesteader had planted. We had a couple of milk cows that eventually grew into a small herd. The milk was kept in big aluminum creamers that had to be washed and scalded every morning and evening.

We had few neighbours. Relatives seldom came to visit. In the summer, it wasn't uncommon for me to go the whole two months without talking to anyone but my family. A few people from the nearby town would come to the lake in the summer, but the long, rutted road discouraged most people. We had our own small and private kingdom with my father as the autocratic ruler who made it all work. When I was a child, I followed him everywhere around the farm. I walked behind him. He was a tall man, with powerful long legs. Proudly, I stretched my legs to match his strides. I did whatever I had to do to match my strength to his.

I used to secretly admire my father's hands. They were huge, the skin thick as leather mitts, littered with nicks and cuts, the lines embedded with grease and dirt and oil. No matter how much he washed, they always had those black lines that smelled of grease and dirt. I thought that was how a man should look and smell.

Together, my father and I fought the enemies of the farm. Every summer, when it was time to harvest the cherry crop, we had to hurry before clouds scuttled over the mountains and rained all over everything, before the cherries split and rotted. We had to get the hay in before it got too dry or too wet and mouldy. Bears broke the apple trees. Coyotes, skunks, ravens and weasels went after the chickens. The contrary cows broke through fences and got out on the road.

Very occasionally, he'd take us fishing. We'd hike to a remote creek across the lake to look for brook trout in the high rushing pools of Next Creek. In the summers we'd go on picnics and in the winters we went skating on the marshy spaces of Rat Slough where he chased us with bulrushes breaking open in a foam of seeds. Or he'd take us out in the rowboat to go swimming and then swim under the boat and grab our legs, pretending to be the lake monster as we screamed in a mixed frenzy of joy and fear.

But despite how hard he and my mother and my siblings and I worked, my father never made any money farming. The chickens got diseases, or the year we planted an acre of tomatoes, the bottom fell out of the tomato market. My parents fought about money, often and bitterly. I dreaded the harsh, loud voices I could hear from my upstairs bedroom—sometimes a chair thudding to the floor, or the back door crashing open when my father stomped outside.

To supplement the farm income, my father drove hundreds of miles to sell milk, meat, eggs and chickens. Or he got a job at the local sawmill loading lumber into boxcars. One year when he was loading lumber, he got pneumonia. He went on working in the rain until he collapsed. My mother put a layer of newspaper on his back and ironed out the pain with an iron she heated on the woodstove.

My father saw any pain or sickness as betrayal. I came home once from rounding up cows on horseback with the neighbour kids with my scalp torn open from a low-hanging branch. When I walked in the door, dripping blood, my mother demanded we drive the twenty miles to the hospital. My father was furious at the idea of wasting gas and time over something so trivial. But my mother won, for once, and a doctor used nine stitches to close the cut.

———

I learned my love of storytelling from my father. I loved to listen to his stories of farming, animals and work. In rural culture, the men have all the best stories. Their language is full of edges and danger. They reshape things as they talk. Among them, the world is dug up and beaten into shape, and things—machines, animals, nature—are fought, defeated, fixed, made to work properly.

I decided to become a writer as soon as I sounded out my first story. I fell in love with the magic of stories and books. They told me things about the world different from what I learned from my father.

In learning to write, I could steal the power of stories away from the men. And I could make stories other than those of the women of our family, who most often spoke to each other in the kitchen, spoke the truths they knew, which were usually all about men, portraying them as large, sullen children.

Learning to read was an easy and logical process. Mrs. Hare, the one teacher at our one-room, seven-grade school, believed firmly in phonics, which meant that we could learn words on our own by sounding them out. Once I figured this out, I worked furiously at learning words. Learning a word was like opening a box within a box, only to find the boxes were all connected together. Eventually, in grade one, I connected enough boxes to read a whole book, *The Little Red Hen*, which I laboriously sounded out, word by word until with delight, I realized I understood the whole thing.

—————

As far as I know, no one in our family had ever met a writer or had any idea how anyone went about being such a thing. We were farmers, or at least, my father was and his father and his father before that—generation upon generation of stolid, stubborn, surly Armstrongs, who'd been farmers and outlaws in Scotland and who, when they came to Canada, married above themselves—women of gentleness and refinement and learning. My father's mother died when he was seven. She was an artist and a musician. She and my grandfather moved from a middle-class life in Ontario to grow wheat in Saskatchewan. I have always been sorry I didn't know her. Perhaps it was because of her heritage that books made a hunger in me that grew and grew even as I fed it and so, like any addict, I went looking for more.

There were books in our house—an assortment of leftovers from previous generations that had been left in the dusty crawl space over the stairs. There was a set of stories of Norse mythology, which I loved, and a *Girl's Own* annual from England, which had somehow migrated to our attic. I didn't understand a lot of what the girls in the story were saying or doing but that didn't matter. I think I was in my forties before I realized that in England a jumper was a sweater and not a pullover kind of dress.

My parents occasionally spent some of their scarce money on books. My father bought us a set of *Science*

Made Simple books once, which none of us read despite his complaining about how much they had cost. I tried but they were dull. But I read my way steadily through everything else the house had to offer, through all the *Reader's Digest* condensed books, which my parents kept subscribing to. I read through the few books in our book-shelf at the school, through our assigned readers past my grade and up to grade seven, which was as high as our one-room school went.

I read on my bed when I was sent upstairs to clean my room. I sat on the edge of the bed and thumped my feet on the floor to make it sound like I was working, until my mother crept up the stairs and caught me. I read under the covers with a flashlight. I read sitting upstairs, shivering in my cold room when I was supposed to be doing homework.

Reading and writing gradually became my secret life. They gave me tools, information and ideas, and opened up new worlds that led me into many other places in my life. Writing gave me the ability to reflect and to shape my sense of self over time. My ability to use language and re-tell my story was important because it separated me from my father's story of who I was or should be and gradually alienated my world from his. This played out in two particular and related areas: in the differences between how we related to land, and also in our deep and angry differences about how I should behave as a woman.

⸺

The open war between my father and me started when I turned into a girl. Turning into a girl was a confusing process. For one thing, I didn't want to be a girl. Girls had no fun. They stayed in the house and did housework. The few girls I knew were silly. They didn't know anything about the things I loved—farming, the woods, horses.

I wanted to be a farmer but how could I be if I was a girl?

More and more often, my father sent me into the house to help my mother. I was supposed to be cooking and cleaning, doing laundry with the wringer washer, washing the milk cans, making butter and jam and canning things and baking—all the work my mother did with barely a pause in her swift pace from morning to night.

In the house with her, I saw my mother's burden and worry. When I was seven, in a sudden desperate fit of tantrum and energy she decided to leave my father. She asked me if she should go and of course, after thinking it over, I said no. How could she look after us, I pointed out to her. How would she manage? But I was never sure I had given her the right advice.

It took me a long time to become my mother's ally against my father.

At first, quite reasonably, I tried to agree with both of them and I tried to get them to agree with each other. I went back and forth during their fights. As I spent more time with her, I took my mother's side more and more often. My mother and I—over the baking, washing, peeling, mixing—agreed genially and often that my father was a bastard and that her marriage to him had been a mistake and as I agreed with her, I turned against my father in anger and bitterness.

—∞∞—

Over many years now, I have seen my father in many different ways. As my understanding of other issues has changed over the years, so has my understanding of my father. The meaning of our relationship would shift depending on from where I viewed him—as his daughter, his ally, or as my mother's daughter, my sibling's sister and later in life, as a feminist, as an ecologist. Of course, none of these points of view are separable from the others, although I can isolate parts of them.

My father lived his life ensconced in his own history, with no wish or desire to change it. So for me, understanding my father has also meant understanding his context, his belief in himself as a man, a belief and value system which was embedded in his history as a Depression-era farm boy, a man who watched his country and contemporaries survive a war, and a man who was a husband and a father in the forties.

He comes from a generation of families who lived on the land and saw their children leave into new and incomprehensible worlds. My father's generation, shaped by war and depression, knew how illusory safety is.

My story is also embedded in a time of social movements, of feminism, environmentalism, and other upheavals, as well as a vast, almost unnoticed and rarely commented on, diaspora of rural people into cities, so that in my lifetime (a relatively brief span of time), Canadian culture has gone from being a largely rural culture to a largely urban culture.

My father gave me a few other gifts. When I was eight, after a lot of whining on my part, he bought me a horse. Then he complained so much about the horse that after a couple of years, I sold her. He also surprised me, when I was nine, by giving me a .22 rifle, although he never showed me how to use it. I figured that out myself.

But I also now realize he gave me other, more subtle gifts. He taught me toughness, to fight impossible odds, to be angry, and above all, to survive. He taught me not to cry, and not to ever give in or give up—not bad traits for a writer.

"You're just like your father," my mother would say when she was really angry with me. It's true. I look like him, I walk like him. When we used to talk together, we talked about farming, using the same phrases, the same intonations. And thus, most importantly, he gave me, unknowingly and unwittingly, my connection to this land that has sustained me throughout my life.

As much as I would have liked to love my father, it seems an anaemic word to describe what passed between us. The attachments that both bound and divided me from my father were tightly wound out of so many varicoloured strands that the simple name of love seems to disconnect it from any meaning at all.

So, in my father's gift of food, I understood his invitation to remain, at least in part, in his world, to understand and share his ideas of thriftiness and survival. I saw the interconnection of both love and tragedy. As a writer, what could I do with such a gift except try to understand and value it, and through that, to understand and value my father.

⸺

Now that I am back living at the farm, I do what I have always done. I garden, I care for the animals, and I go for long walks and listen for stories to come back to me from the stones, the sky, the dark lake water. I search among these stories for truth the way my dog noses among the dried bracken and knapweed, the way he watches intently for ground squirrels when we go to the beach. I watch the mountains—he watches the rocks. The dog reads the ground for stories the way I read the people and events around me.

I will never get to the bottom of the layers of truth about my father, any more than I will get to the bottom of the layers of truth about myself, but in many ways, I realize now, I am my father, so much so that I have the temerity to think that I can stand in his shoes and sing his song for him. I will always share this land with my father, the greatest gift he could have ever given me. I live in his house. I wear his old coat and hat when I do chores. I wander the same trails I walked as a child. I walk between my father's world and my own, and in that dissonant space between a history that is being lost and one that is being created, my father and I and the land continue.

Sad Songs

For too much of January, I felt like bad old, sad old country song. My old dog died, and then two weeks later my mother died. I sat beside her, held her hand, and tried to let go. It was a deep dark miserable January night, raining hard.

It was hard to watch my dog get old and struggle with keeping his dog dignity, running, barking, peeing and shitting outside. How do dogs understand old age anyway?

All fall, I went to see my mom as much as I could manage, trying to be with her and watching her decline and decline. In some odd way, her death has given her back to me, my lovely, laughing, always busy mother, who was never quite the shrunken white-haired woman in the wheelchair watching the door for me to come in.

Tonight I am again, finally, alone, funeral over, family gone, rain on the roof, only one dog snoozing on the rug. It is unseasonably warm, and I am ignoring the whispers from the greenhouse, the trees and plants. My brother was out pruning trees briefly today but we both know in our bones it is too early to pay any attention to the garden, even when the lying sun pokes out from behind the clouds with actual warmth in it. It isn't warmth but light that triggers growth.

When I am outside on a cold spring evening and I can smell the ground, a smell of cold and mould and earth and anticipation, this tells me it is time to wake up and get ready for spring.

But right now, the light is still telling me, sleep. After the funeral was over, after the house was clean and quiet, after a week of decisions and dinners and the jangling and jostling and banging of various people, emotions, sensibilities and ideas, I curled under a heap of blankets back in my own bed and let it all go, the dog, my mother, the family wrangling over funeral arrangements.

It always surprises me when people ask if I live alone. Tonight, the house is crowded. My father's chair in the basement creaks when he sits down. My mother sits beside me, glass of home-made wine in hand.

Farmer

THE YEAR AFTER OUR FATHER died, my brother (with whom I now shared the farm) and I had to come to terms with killing animals. There was only him now, to do the killing and butchering. We had raised a bunch of pigs. We've always liked pigs. They're smart and funny and we always get attached to them. We take them for walks, scratch their ears, give them names. They are also easy to feed and delicious to eat.

The day my brother decided to butcher was pouring rain. It was late November. He'd been putting it off and now it was getting late in the season. He's a logging contractor in his working life, and he loves the farm. But it's hard for him to make enough time for it all.

He fed the first pig a mixture of grain and beer, thinking this would calm the pig down and make it sleepy. He would shoot it in its sleep. Somehow that might make it easier. But the beer made the pig happy. I went over to the barn to see how things were progressing and there was my tall silent brother standing ankle deep in the mud, head bare to the pissing rain, while the pig galloped clumsily, mouth open and looking happy, from one end of the pen to the other. I left and when I got back to the house, I heard the shot. I went back out but there wasn't much I could do to help. My brother skinned out the pig and threw the legs to the dogs and loaded the carcass in his truck to take to the butcher. It took him about two hours and then he came in the house to get warm and dry off. I made the tea while he sat silent at the table that was once our parents' and our grandparents'.

Finally my brother, who never says much, said, "Dad was so fast." And then he said, "I like to be alone to do this." All of which I understood. The subtext was there, missing our father and the feeling of being alone and grown up and responsible adults, even at fifty, along with the harshness and the duty and the beauty of killing the pig, which meant good meat for winter, safe healthy meat, work we could feel good about. We exchanged some small talk about the butcher and when the meat would be ready and then he left.

I am walking to the barn, a daily ritual. It's winter now, so twice a day, I carry grain and warm water to the chickens and throw hay to the cows. In the late dusk, after chores, I go for a final wander with the dogs. When I come inside to warmth and peace, the dogs digesting their dinner while snoozing on the rug, I have a satisfying sense of the farm settling to peace with me.

I have always thought of myself as a "farmer," but this is a self-chosen identity and doesn't have that much to do with how I make a living. Labels about things to do with land are various. My landscape architect daughter says she is a gardener; my neighbourhood friend K. Linda Kivi calls herself a peasant. It's all vaguely about mucking about in dirt and playing with plants. I call myself a farmer and am content, and my son laughs at me for thinking I have a farm, and he's right. It's not much of a farm, really, no big machines or agribusiness connections, just a bit of a clear cut beside Kootenay Lake.

The first white man who lived here and owned the land was serious. He ploughed and planted trees; he ditched and drained and built a house, a barn, a chicken shed, a woodshed; and he fenced and cross-fenced the whole place. I have tried to imagine him but I know little about him other than that I still benefit from his work.

This place, even if it isn't a "real" farm, according to my son, has fed our whole family—my father, my mother, my three siblings and me, and then my children and my brother's and my sister's children as well—for over fifty years. It is a lot of work, and I did much of that work when I was a child and then when I was a single parent with four kids to feed. So I feel I have earned the label.

~ooo~

Farmer is a word with odd implications. It is one of those words, like "Indian," with romantic stereotypes attached to it and a similar paradox of stereotypes—on the one hand, a farmer is someone often perceived as being close to nature, wise in the ways of animals and plants, someone who can fix anything, do anything, as well as encapsulate a mystical view of the universe in just a few well-chosen words. On the other hand, the word "dumb"—as in "redneck," "slow," "prejudiced"—is often silently appended to the word farmer. As well as that, farmers are almost exclusively considered to be male.

The other odd thing about farming and gardening is that as long as you do it as a hobby, you can be a "gentleman" farmer, which is different from being a real farmer because you have money, or you can be a "hobby" farmer, meaning you can have your place in the country where you garden obsessively and even keep a few happy cows and be indulged by society in your eccentricity.

But a real farmer is forever, no matter how wealthy, in the lowest class one can be in. It seems to me people don't want to be farmers anymore and that western civilization has been a lot about crawling out of the rural peasant class and into something more respectable. Becoming a farmer means going down a long notch in the class war system. Despite the romance and the beauty and the great food and other benefits, being a farmer has a very old image problem. Somehow, it seems manure on the boots translates into manure in the brain.

There is, I think, a farmer's walk: slow, bent, but steady. Farming is walking, working, stooping, lifting. It is very physical. As well, there is thinking, observing, deciding what needs doing and in what order. There is no point in rushing. The trick in farm work is to keep going at a slow and steady pace that will carry you through from morning to night.

As I get older, I have become more like my father and probably much like his father and his father and his father. Gender doesn't much enter into it. Superficially, in every way, I am utterly unlike them: I am female, a writer, an academic, an environmentalist, a feminist; and yet none of this really matters on the way to the barn, or wandering the yard, checking things, gates, tools—the ritual and the rule, as old as humans, bred in my bones: put the farm to bed before coming in for the night.

It always amuses me that prostitution gets labelled the oldest profession when the obvious oldest profession is farming. Farming is about food and survival, any romance about it is just that, furbelows and frills added onto the substantive nature of the task that is really only about feeding yourself, your family or clan and thus staying alive.

Farming is fundamental. It is about food. Thus it is also about dirt, physical hard work, being outside in every kind of weather and, of course, birth and death. It is also relentless. Farming never stops, it just slows down in winter. There's nothing that can be put on hold—fruit can be green one day, ripe the next and rotten the day after, depending on weather, heat, water, soil temperature and light. Animals always have to be cared for first, and keeping animals means that you get to love and care for them and then you have to kill them.

My father loved his animals although he never let us see that. We only knew it accidentally, by coming on him crying when a calf died, or when he had to shoot an old dying dog. But he was fast and skilled at butchering. When I was a child it was an exciting time, killing things. I never connected the fall butchering time with death. To me it was about work and food and curiosity, about the gleaming mysterious insides of cows and pigs, and watching my all-powerful father dismember the carcass. I think I was probably a teenager before I stopped being fascinated with all this and began to avoid it, not with disgust, but with confusion. I wasn't sure how I felt anymore about killing things. I was amazed that an animal could change so quickly, from something alive to something that was now food, meat. The change was incomprehensible. I stopped watching.

My work at the farm is primarily growing vegetables and fruit, which is, I admit, traditionally women's work but I don't much care. Let my brother have his machines while I obsess over cucumbers and beans. I grow flowers but they don't get my attention like the vegetable garden does. Every year, the vegetable garden is a journey into adventure, very small adventures, to be sure, but always interesting. One year we were invaded by voles, small brown furry creatures with enormous appetites. Since we had never had voles before, it took a while before I began to catch on to why the tomatoes were disappearing and the eggplants were hollow and the quinoa stalks were chewed to stubs. I couldn't even decide how to react—should I kill them all, wait for winter to kill them, put up with them? In the end, the voles and I shared the garden. They ate half and I got the other half and then winter came and ate them up. Or something did.

What I like about growing food is texture, colour, shape, the sense of abundance, the sensuousness of it all; for example, the way the late August light glints off a basket of pink and purple eggplants, red and green peppers and yellow and orange tomatoes. I revel in the satisfaction of picking things, but then I try to find someone else to help can, pickle, dry, bake, juice and freeze the summer harvest. I like this process as well, but mostly I like the sense of community and shared effort when there's a bunch of us around the table, peeling peaches and slicing them onto trays for the dryer.

Because small-scale agriculture is mostly tedious physical work, the more people there are, the more food can be produced. A small farm can feed a lot of people easily but all the work is intense and the adage of many hands make light work is exactly right.

But this means a small subsistence farm is also about community, whether you want it to be or not. It's about geographical community and it's about family; you don't choose who you're related to or who moves in next door to be your neighbour. If you all live there and you have to do the work to feed yourselves, you have to get along. Or leave. Small farming is about depending on your family and neighbours. It's at once both oddly dependent and a powerful source of independence from systems, corporations, governments and grocery stores.

Industrial agriculture replaced that sense of community and freed us from all that neighbourliness and getting along, through mechanization and industrialization. It freed people from the stigma of being farmers.

Consequently, these days, most people who live rurally aren't farmers, they're either middle class people who have retired, or people who have bought a second—or third—home in the country and then sometimes even gated it off. Instead of being the place where people work, the country is now where people play. Because people have moved from urban areas where often neighbours are just people who live in the same geographic area, rural community has mostly lost its traditional ethic of "neighbourliness."

But with the rising price of everything—oil, energy, food and land—there is in my part of the world, a quiet but insistent buzz about farming and whether small farms, subsistence farming, even local agriculture, might make a comeback. On a broader scale, in parts of North America, the new movement has all kinds of recombinant parts and labels: "re-localization," "the one-hundred mile diet," "local," "organic" food.

Those of us who were part of the back-to-the-land movement in the seventies are trying not to be too cynical. And people like me, who were never anything but local, are waiting to see if this actually translates into more money for farmers.

Farming is complex but it's also simple. Farming is basically about things growing and since things really want to grow, a lot of farming happens almost accidentally. Seeds in fertile ground will do their best to grow and reproduce—given enough water, warmth and light and not too much competition, they'll do just that. Animals want to have babies and reproduce, and there's always too many male animals and nothing else to do but eat them. And of course to do that, someone has to kill them. There's also a lot of work involved in eating meat—things have to be butchered, preserved, put away for later.

If you're not trying to make money or produce picture-perfect fruit, farming can be a story of amazing abundance. But of course, it is also a story of weather, animals, weeds, insects, disease and frustration. Perhaps that is why my father saw the farm as his enemy, surrounded by other enemies. He was depending on the place for income and survival. I'm not, at least not at the moment. Would I see things differently if my family and I needed whatever we grew to survive, and if weather, deer, coyotes, bears, were a threat to that?

—∞—

So why do I call myself a farmer and why have I stuck to this label so fiercely? I have asked myself this question for years with no real answer. I have asked myself this not only walking to the barn in winter, but standing in the muddy freshly turned garden soil in early April, planting seeds, or in early June, stuffing my mouth full of new spinach, or wandering among the peach trees, bucket in hand, looking for the first, earliest, just-getting-ready peaches, or in the fall, putting the garden away, pulling onions, cutting cornstalks, with the chill breeze rattling the leaves.

My complicity here is that I don't identify as "just" a farmer. I have all those other labels to fall back on, depending on where I am and to whom I'm talking. But farmer is a word that tugs at my heart, or perhaps my feet. It's there among the fundamentals, digging manure out of the pigpen, for example, that I find my own peculiar sense of freedom within the boundaries of belonging, of connection, and of work that is fundamental because it is just that, work that feeds me and my family.

I am well aware that all of this right now is an intellectual luxury, that even though I don't have enough money, I have enough to buy packaged food if I need to, and on nights when I am feeling lazy, I can go to the local pub for pizza. And I am also aware that for several thousand years, it wasn't a luxury, that small farming was the way generation after generation of people survived, that small farmers were often an independent and hard-headed lot, as long as they clung to the land, separate from politics, armies, and the changing fortunes of whatever kind of nation-state they happened to be in.

Perhaps that's the connection. That's why, as a writer, I cling to this word, this label, this old-fashioned, out-moded, roots-in-antiquity lifestyle. It's that hard-headed independence, that sense of being outside, of not having to care about the latest trend or fad, whatever it might be, being outside of history, even outside of time itself. It's the ultimate in arrogance, in fact. Being a farmer enables me to think what I want and to live how I want and to let my feet come down on the earth in tune with some ancient rhythm—even if I am only dimly conscious of it, a rhythm laid down by my ancestors, my parents, the land itself.

Money Land

All month, I have been dealing with the farm as a piece of real estate, a dichotomy that drives me really, really crazy.

Today was sunny. The heavy grey sky lid that sits over the lake day after day and screens off the mountains burned off by noon, and I went outside, pulled my baby maple trees out of the sawdust and watered them, planted some sprouting apricot pits and went for a walk.

We made a number of decisions about the farm, most notably to invite a couple of our friends to put RVs on sites near the beach—a break with our traditional family culture of proud solitude—but we did it because a farm endlessly needs money and infrastructure, fences, barns, pens, tractors, seeds, mowers and so on. It needs to pay its own taxes.

We also contacted a lawyer and real-estate agent and met with both of them in order to plan a small subdivision. We met with the bank loan officer and I drafted a "business plan."

This process will go on for quite a while. I loathe it, loathe the concept of land as real estate. Want to hide from it all. I want to leave the farm untouched and intact, even to let it go back to wilderness as much as possible. Instead, inch by inch, we are surrendering to civilization. To roads. To people. To buildings.

Everything we are doing makes sense and it's all for the best of reasons. The people with RVs are people who would be visiting and sharing the beach anyway. Our friends.

The lawyer and the real-estate person are both good people. But it all makes me crazy. This land is not real estate. It's my home, a place I want to protect and care for, a place that protects and cares for me. Not as real estate, not as private property.

I am not a person who likes or wants to deal with money, real estate, wills, taxes, contracts or any kind of paperwork. For me, in this shadow world, where contradiction lurks in every corner, values shift and change, and the future of the farm and the nature of my connection with it suddenly seem threatened and fragile.

Mine: Notes Toward a Land Ethic

I WAS WALKING BACK HOME to the farmhouse from the beach one night quite a few years ago. I like to go for a walk around the farm just before I go to bed. My excuse is that it's a farmer's habit, to check on things, but really I just like to wander around in the dark, the dogs running beside me, and the night sounds all around. But this night I stopped. I bent over and touched the ground with my hand. "Mine," I muttered. "Mine."

And even while I did it, I wondered what I was doing. My father was aging and alone. My mother was in a care home with dementia. My father refused to make a will. The future of the farm was in doubt. But my younger brother and I had just made an agreement to try to keep it, to try to maintain the life we had lived since we were kids. It wasn't greed that was in that statement or even possession. It was something deeper and older, something I am still deciphering.

———

I have thought about that moment many times, especially after my father died and my brother and I now "own" this land, pay taxes on it, take care of the gardens and fruit trees and houses and animals and other things that have become part of it and part of us in the last sixty years. We are making plans to re-orchard, replant, rebuild.

We also live here, at least I do, and my brother comes every day to do the bulk of the work. We have strange elliptical conversations about weather and animals. We rarely sit down and really talk. Like me, he likes to walk around. He spends a lot of time walking up to wild animals, to elk or birds and he keeps me informed of what, or who, is living around us.

But mostly what my brother and I do is what we have always done: we feed the animals, prune the trees, pick fruit, preserve food; and we look, we smell, we notice the light on the mountains, the movement of the clouds, the distant wind rolling through the treetops. We notice these things in their beauty, but also because they are part of us, as we, through our long existence here, are part of them.

When I am away from this place, my existence is truncated and limited by the loss of this noticing. The farm and the lake and the mountains always feel, not a part of me, but as if my consciousness is in some way made by them and by our long interbeingness—or connection. ("Interbeing" is a Buddhist word describing the interrelatedness of all things.)

But that relationship has changed, subtly but surely. So now I need to think hard about what I'm doing here because that odd moment in the dark has come true, and I have no idea what it means to actually "own" land, and I both fear and am amazed at the power in that word, "mine."

I fear this word for a bunch of reasons. I fear the history packed into it and I fear the kinds of ideologies that are attached to it. I hate the fact that it takes me into stories of money and owning and real estate. To me, the word reeks of wars between settlers and indigenous people, of ongoing colonialization and resource exploitation, of different value systems and ways of being trying to dominate each other. It's a bigger word than it first appears.

I have a friend who used to lecture me about something called the "owning" class, to which, apparently, I now belong. In her philosophy, the owning class is problematic in its rich and aristocratic privilege.

But there are different kinds of ownership, aren't there? Or am I fooling myself? Have I now joined the ranks of a class of "landowners" with whom I have never identified? In fact, have actively detested? I think I have always wanted to own this land from the simplest and clearest of motives: because I love it, because I identify with it, because I am at home here. But can owning something ever be that simple? What motivates other types of owners? Are we all the same in our owning? Can there ever be a pure motive?

—∞—

At the very heart of our current desperately troubled and dysfunctional relationship with the ecology of this world is the idea—and ideology—of land ownership, an idea that is rarely, if ever, questioned.

We never really look at how ownership conflicts with other shared rights, or the rights of the "commons," because that would force us to question some of our most deeply held and (I use the word advisedly) "sacred" values about land and resources. These values are bound up with the idea of property and wealth as a kind of freedom. John Stuart Mill, in particular, saw owning property and exploiting privately owned resources as a defence against tyranny, and most people in North American today would probably agree with him. Owning things makes you free in the sense that it defends you against poverty and gives you a place to put your hand on the ground and say "mine."

However, John Stuart Mill never envisioned global warming, mass-market consumerism, environmental destruction on a global scale, and corporations bent on making fortunes through destroying the environment.

And although we have ethics and guidelines about how to live with and care for other human beings, as it stands now our laws and western culture don't recognize, or even consider, that we must live on better terms with the natural world.

Our laws about land and ownership are based on the crucial assumptions that a) we are capable of owning parts of this world, b) our own needs as humans override any other need that might exist, and c) that we are, as humans, innately superior in intelligence to other beings on the planet. None of the people who formulated these ideas foresaw resource depletion, species extinction and environmental collapse.

Given our present ecological mess, we need a broader discussion of how to live with the non-human world and how to live ecologically—something that people are discovering is very hard to do. None of the systems upon which we presently depend are very ecological, and to leave those systems means, essentially, making life a lot harder for ourselves.

Now, since I am a landowner, a complex relationship that came about through inheritance and chance, I feel I had better try to understand what I'm doing here. I don't feel like a landowner. In fact, I feel more owned than owner. Certainly, the swallows shitting on my deck, the deer who chew my fruit trees, the bear that devours the plums every fall don't care what I call myself.

But who and what I am and whatever identity I have is inextricably mixed with my relationship with this place. The relationship is complex. The land has taught me most of what I know, or to put it in properly human-centric terms: I have had the sense and good fortune to learn from it. It has shown me beauty, or at least what I was willing and able to see. My relationship with it continues to demand that I question such things as the nature of land ownership.

After all, trying to own land is absurd. It is like trying to own time. The land has been here as long as this earth has been here, and it will be here in some form as long as the earth exists. So what I own is not the land itself but a certain set of rights granted to me by the government, which include such things as the right of tenure, the right of inheritance, the right to pay taxes and the right to change and manipulate the land as I see fit, within certain limits. The government actually "owns" the land. They just lend it to me.

It often astonishes me how lax the limits are of what I can do. I can shoot wild animals on "my property." I can chop down trees, spray any plants I want with herbicide, dig up the ground and even pave the whole thing if I so desire—except for the creek, which apparently does have some protective laws attached to it. There are limits to what I can do to water and limits on how I must dispose of sewage. But other than that I am free to alter and destroy the ecology of whatever plants and animals lived on this land before I appeared, for whatever reason I deem necessary: for money, for vanity, for convenience, for whim. If a coyote or bear or hawk or any animal becomes a nuisance or interferes with my safety or the safety of "my" domestic animals, I can kill it, regardless of how endangered it may be.

Now, if I am a somewhat reasonable, caring and conscious human being with a slim amount of ecological awareness and understanding, I might hesitate before doing any of these things—and I do. In fact, I've found that much of my behaviour and attitudes toward the notion of "stewardship" and ecology have been shaped in direct opposition to the ideas of my father, who "owned" and depended on this land until he died. And although he lived on it for over fifty years, my father would have snorted contempt at the idea of loving land. When I tried to tell him I thought we needed the land to stay in the family, he accused me of wanting to put him off the land. He only made a will because he knew he was dying and even then, reluctantly. Although his farm was and is a valuable piece of real estate, he considered it, as he told us often, a "millstone around my neck."

I understand this far more now than I did then. When I walk out the door, what I see is a million small things that need doing. When I leave, I have to find a house-sitter. I am chained to this place by love and habit and need.

My father lived almost his entire life as a small farmer and he cared about farming as his work. It was all he knew and what he knew he had learned from his father. I have a few times since his death tried to figure out how my father actually felt about this place. For him, it was a place full of enemies: coyotes, hawks, ravens, and especially weeds—and for him, almost all plants, except those he wanted to grow, were weeds. They were "taking over," he would say, rushing out to spray herbicide on the buttercup growing along the creek in the pasture or the knapweed growing along the gravelly roadside, the patch of ox-eye daisy on the path to the beach. For him, farming was war. And yet, when he was lying in his hospital bed dying, all he wanted, he said over and over, was to be able to go out and work in his garden.

A farmer is definitely a person who both works with and against nature. A farm is a clear-cut in the landscape; whatever was originally there is gone. The farmer puts animals in pens, digs up the ground, plants trees in rows and prunes them, lets some plants grow and kills others.

But with a good farmer, the animals in pens are fed, safe from non-human predators, and cared for. The plants, in rows, in their dug-up ground produce astonishing amounts of food, and any smart gardener covers as much of the bare ground as she can with compost, mulch or a cover crop. Fruit trees do well in rows and have to be pruned, or being fruit trees, they set out to grow themselves to death. So the good farmer works within constraints that form an odd kind of partnership with some natural forces while ignoring or actively fighting against others.

The other economic alternative for land is to use it up: cut down the trees, catch the fish, dig up the minerals—or whatever resource there is. In other words, commodify whatever's there—even beauty.

Most of all, land gets turned into "real estate." It is bought and sold and everything about it becomes a commodity. It is bought by people who then begin to build a relationship to the place they now "own"; although by the time they "own" it, whatever existed of the original ecology is long gone.

And yet, of course, many people living on "their" land, even though they never see the ecology and the creatures that are gone, still attach great emotional significance to where they're living. And this emotional significance mixes in with the economic significance of land as real estate to make a murky emotional soup.

When we were children, because we lived far out of town, my siblings and I had a private kingdom. Growing up on this land afforded us a tremendous sense of independence. Despite the relentless workload, we felt free. We had real wilderness to escape to, an endless panorama of trees, mountains and wild animals. We made a kinship out of it all; we knew every trail and tree and cliff. And we translated this kinship into a kind of inverse snobbery, as well as a sense of clannishness and solidarity with each other, even when we didn't like each other much.

And above all, we had the lake, the incontestably, inexhaustibly thrilling lakeshore, where we launched ourselves into adventure, paddling out from shore in the icy spring waters on logs with sticks for paddles, turning stumps and rocks into spaceships; or, in spring, spending hours building dams and ponds in the mud left behind by low water, and almost every day in the summer, after the farm work was done, we went fishing. There is something magically hypnotic about sitting on a rock watching a red and white bobber bob about on the surface of the water. Something might happen, something could happen, and fairly often it did: the bobber jerked or went under or started to move and one of us would reel in a silver, thrashing prize, sometimes a trout, sometimes a coarse fish, a squawfish, or a chub, but always a thrill.

—∞—

When my father died, my brother and I didn't even really discuss what needed doing, we both knew what we wanted. We wanted the life we loved and knew so well to continue. We wanted the land, and the freedom and independence that came with it, along with the work.

To do this, we needed to sell some land, so with some private grief, we decided to sell some of our lakeshore so we could keep our arable land and keep farming.

But of course this meant turning some of our land into "real estate." Interestingly, as I write this book, six years after my father's death, we still haven't succeeded in doing this. We can barely think about it. My brother and sister and all of us sense the idea of selling even a corner lot as a huge loss, like dismembering a beloved body, a parent.

—∞—

Private property is an invention, a convention, an artificial construct upon which most of the world's economy is now based. Although it was an almost unknown idea through most of human history, it has attained an almost mythic status, a kind of untouchable prominence in European and North American culture since the 1600s when John Locke declared it a fundamental right and a bulwark of human liberty against tyranny. It was also a fundamental tool in the expansion of colonialism, since the colonizers assumed that the indigenous people who didn't have such an "important" concept were inferior and therefore their land could be taken away from them and declared "owned."

Today I walked down the shared road between our property and our neighbour's. He has sprayed his side of the road with herbicide. He knows we're organic farmers but it is his property. He has sprayed the creek banks as well. He thinks Roundup is safe. He told me last time he saw me that Roundup is safe enough to drink. I should try it, he said.

Recently our new neighbours to the north spent weeks and a small fortune drilling and blasting into solid granite rock in order to make a boat harbour. They are no longer allowed to build one out into the lake, so they spent a lot of money to make a hole in a cliff in which to park their boat.

Some other new summer people built an enormous summer home, projecting out over the water. They put up a sign naming the place "Wood Nymph Trail." Their lot is a piece of granite bedrock, covered with stunted firs, which they drilled and blasted for a month to make a level place for the house. In the Ktunaxa story there are powerful and ancient spirits here, and then there's the lake, itself a presence, muttering in its deep narrow bed, turbulent, wind driven and cold. It's about as un-wood-nymphy as a place can be.

Today it was raining and I couldn't work in the garden so the dogs and I hiked up the mountain. I haven't been up there for a while. From high up on the mountain, our place is a ragged patch of green, a tiny clear-cut on the toes of a huge mountain beside a vast dark lake. Never mind that the mountains above me are wound about with roads, that the lake is dammed and controlled. It still has many ways of reminding me of my own insignificance. From high on the mountain, our farm seems small, a little patch of a place.

When I came back down, I sat on the porch for a while. I try to spend a period of time every day just listening to the endless conversations around me; it's like trying to learn a foreign language by sitting in a café beside a busy street. There are birds and gophers and crickets, frogs and dogs and chickens; there's grass and flowers and bushes and vines. Beside the driveway is a tangle of thimbleberry, maple seedlings, blackberry vines, burdock and dandelions all fighting for light and ground. Sometimes I feel like it is all shouting at me and I can't hear a thing.

None of this has much to do with ownership. It has to do with meaning. Where it gets tangled up with ownership is in the question of access, of rights, of privacy. I could find all this peace and beauty in a park, but then I'd have to share it and it wouldn't feel the same.

But why not?

I love to share this place. I love people to come, visitors, relatives, whole parties of people. I hold a summer writing retreat every year. But of course, I get to choose who comes. I get to be in control. Being in "public," being outside those boundaries, means losing control, being just another face in the crowd. I may believe in love and peace, but like many people, I don't like other humans I don't know in numbers that I can't control.

When my visitors go away, I am alone in the peace and the silence. I love my friends but I'm always glad when they're gone.

Within our homes, our place within that sense of "our," is the ability to stretch out, to be eccentric, to sleep, to fart, to eat too much and throw newspapers on the floor and junk in a corner. Within that larger space of land is the ability to be, paradoxically, eccentrically, fully human. Could I do this in a rented apartment in Vancouver? Sure. Would it feel the same? No. What makes the difference? That little word, "mine," with all the centuries of privilege and power contained within it

How would I change this even if I could?

——

Not long ago, I read about a conflict a landowner was having with his neighbours. He had recently bought a piece of land through which ran a much-used trail to a public beach. He gated it off and the local community was outraged. His reasons made sense. He wanted to protect his land from garbage, from people trampling gardens, leaving open gates and scaring his animals. Nor did he want to be sued if someone fell down. The community was equally reasonable. They had always used the trail, there had never been any problems and they wanted to go to the beach. I agreed with both of them, but in that landowner's place, I'd have done the same thing.

It's not about rights. It's about ethics, behaviour, cultural values and understanding and knowledge of ecology. A group of people walking along a trail who know nothing about where they are, where they are going or how to behave is one thing. A group of people who share and respect a land ethic, an ability to be quiet and listen, is another. The second will care for the place, but the first won't. Unfortunately, there is no way to tell which person belongs to which group.

So when I invite people, I assume they will listen to what I tell them about bears and garbage, about fires, about sudden storms on the lake and small boats. But over the years, other people have come uninvited. They are the ones who leave bags of garbage and beer bottles, who once chopped a wooden ladder for their bonfire, and who drove into the garden beside the blackberry patch one day and began picking berries. When I, bemused, asked them what they were doing and if they knew they were on private land, they said, "Well, there was no fence."

Since the only shared public, cultural ethic most people today have about land is that of exploitation and destruction or, conversely, of consumption and sentimentalization, then how can we expect people to behave? Culturally, with respect to land, we have no manners. We're like two-year-olds dressed up and turned loose at a party. Ownership is apparently both a barrier to the yobs and the right to act like one.

When I was growing up in my rural community, it was a place of farmers, miners and loggers. The ethic was one of sharing, or rural neighbourliness, based on the idea of pioneer survival. People gave what they had: they traded labour, tools, garden produce, babysitting, rides into town. Paradoxically, when the back-to-the-landers, looking for that very thing—community, arrived in the early seventies, the community began to change. The new people didn't see what they were looking for because they came with urban eyes, funny clothes and the idea that they could remake country life to suit their own values. And all along, the vacationers, the summer people, the cottage owners continued to come.

In the last couple of years, the community has begun to change again. Rural land prices, especially waterfront, are rising fast and now we are getting monster houses built by the rich and then gated off. Community appears to be the last thing these people want. But of course, oddly enough, I am now in some sense one of them, one of the big land-owners, contemplating gating off our place to keep out the day-trippers. And yet, though I feel I have nothing in common with my neighbours because I acquired the land through inheritance, rather than riches, if I look like them, and act like them...

⸺

On my walks now, I often think that there is so much that is detestable about modern civilization, so little that I like or approve. Some days I think I am an anachronism; some days I am the future; some days I am just a grumpy old woman complaining about change.

And all of this begs the question of how and what I would change if I could. Would I give up "my" land to a kind of communal ownership? Could we make our own village here? Turn back time or, conversely, leap into another kind of future?

I've considered this. There are many days when it's lonely here in an odd and particular way. Harvest should be a community celebration time, singing time, bringing in the sheaves time, continuity time. But usually it's just me. I love picking garden food, peaches, plums, apples, but it's a lot of work, and I wouldn't mind a crowd around to help. Not a big crowd, just a few like-minded people who love the work and love what it means and love farming and love the land. And don't see it as property. As real estate. Don't think they own it. Are just glad they live here and glad the land gives them sustenance and beauty and rituals. With those people, I'd share.

Slowly, this is happening and I am becoming less cynical about it all. The growing awareness of the importance of local food has a lot to do with this. Plus, I share the land with my family, my children who come and go from the city. Often their visits are oddly wistful. They love the place. They like coming to visit. They couldn't imagine living here. They have busy fulfilling lives and careers. One of my daughters is a film reviewer. These days she watches too many documentaries about the ends of things: the end of food, the end of weather, the end of glaciers, the end of water. She phones me, worried.

"Here I am," I say. "The farm is here."

———

One night, again, walking around, I paused, thinking about the conversations with my daughter, along with the thirty years of Kootenay conversations about what we'll do when the world does fall apart. If this place meant survival for my grandchildren, would I fight for it? Yes, I would, and that thought really surprised me, being a non-patriotic, pacifistic kind of person. It felt old, this feeling, and not about ownership at all, but about something more basic, the animal knowledge of place and the right to live and survive there.

But there it is again. The contradiction. Fighting over land? Would I fight then, with the bears over apples in the fall? Would I fight, like my father, with the weeds in the garden and the coyotes who want to kill my chickens?

No, more than anything else in my life, I want to live here in peace, share the land with the other beings that live here—wasps, bats, swallows, flies, ants, snakes, coyotes, bears, cougars, white-tailed deer, eagles, ravens, ospreys, geese, and yes, even my neighbours. I want to understand how to do that, even if such a process is full of contradictions, even if it seems impossible, even if there are no ethical standards to guide me. That means sharing; it means coming to some kind of understanding with the people who want to walk over my land and throw garbage on it because they see it as empty, or perhaps they see it as a park where someone else will pick up the garbage, or perhaps they don't see it at all.

It is impossibly conflicted and difficult and just for that reason, all the more important. Although I have somehow acquired the power to dig it up and chop it up and cut down trees and plant other trees and kill things that live here, this land isn't a "resource." But I am constrained by my own understanding, which is still limited, and my own values, which are still developing, and by my own mystification at the emotional, intellectual and communal perplexities into which I have now been thrust by that simple word, that act of stooping and acknowledgement, of saying "mine."

Inside and Out

The first swallows have arrived, dive-bombing each other, playing in aerial acrobatics. These swallows will move on north. The next wave will stay, build nests all over the porch, raise many babies, and in the early mornings as I drink my coffee on the deck, they will sit on the log over my head and gabble swallow talk.

The frogs started early as well. Now they are a brilliant chorus that punctuates and swells and billows through the cold spring nights.

We are early with everything this year. It was a dry warm winter and will undoubtedly be a dry summer. Already the gardens at the farm are dug and ready to be planted. Several people are gardening here this summer, which takes the burden off me and makes the work so much easier.

It's a mixed-up month: storms and squalls barrel across the lake, huff and puff, blow away, the sun comes out, it's hot, oh no, it's cold again. It's impossible to put on the right amount of clothes. I don't need a fire and then I do. It's too early to plant the garden and then, maybe it isn't. Maybe the onions can go out, or the broccoli, or maybe wait.

Easter is next week. I have two hens setting so perhaps baby chicks will appear if the eagles, weasels, skunks or coyotes don't get them first.

It always amuses me somewhat that baby chickens and rabbits are symbols of Easter. No food in the garden yet, no

*fruit on the trees and most of the last year's storage exhausted,
what would there have been to eat? Easter is about starva-
tion foods—eggs and rabbits, and perhaps, if you were really
lucky, a lamb.*

*My friend Kuya told me yesterday me that the Buddha
said the greatest human illusion is that our bodies end at our
skins. From my view on the deck, I get to watch not only the
swallows, but the eagle on the pine tree every morning, the
slow emergence of buds and seeds, spring again inside and out.*

Interbeing/Animalia

I BEND OVER. I PUT a plank up to the door for a ramp. I pull open the small door. A few chicks peep out, stare at the light. They crowd around the opening. I sprinkle bits of grain on the ramp. One ventures out, then another and another. They step on grass for the first time, lifting their clawed toes. They peck at leaves, grass blades. The sun shines hot on my head. I crouch in the grass. The chicks peck at my toes and hands. I am seven. I am immensely happy.

The eminent biologist E.O. Wilson has suggested that children naturally are attracted to other non-human living things, a trait he termed biophilia.[1] When my grandkids come to the farm, what they want to see are animals, the pigs, the chickens, the baby swallows. They feed treats to the cat and the dogs.

When I was a child, I lived far more intensely with animals, both tame and wild, than with people. There were always a lot of animals at the farm, and my brothers and sister and I made pets of all of them, every calf and pig and dog and barn kitten. We also brought home fish, turtles and sad baby birds that always died.

The only things that weren't pets were the chickens. There were simply too many of them. But one of my favourite jobs was to care for the hundreds of baby chicks that we ordered every year. They came, cheeping and thirsty, in shallow cardboard boxes. One by one, I picked them up, showed them the water and grain in their new home, the floor spread with clean sawdust. They huddled together under a metal hood, where a glowing red sun lamp mimicked the warmth of their lost mothers. I fussed over them. If they huddled together too much, they'd smother. If they were chilled, they'd get sick. But usually they thrived and then one day, I opened the door to the big world, a pen full of green grass and sun, and watched as one by one, cautious and fearful, they ventured outside.

Eventually, the hens went off to the big chicken house and the roosters went into the freezer and I lost interest in them. There were so many other animals. Late one rainy spring night, my father came home. He called us down-stairs, brought his hat, full of wild baby mallard ducks, out from under his jacket. The mother had been killed on the road. We kept them in the house in a box behind the woodstove, and when summer came, we made them a house outside with a pond in the creek. They followed us everywhere on the farm, walking until they could fly. They followed us to the beach all summer, flying just over our heads, and when fall came, they sat on the front lawn and honked mournfully at other passing ducks and geese. And then one frosty morning, they were gone. They came back in the spring, but then they were wild.

They, or their descendants, nested in our pasture. The fish, turtles and frogs went into the small pond we had made beside the house. They always escaped. We didn't mourn them. There were more.

⟨⟩

At night, I curled under the quilt in my upstairs room
and read books about animals. Stories about horses were
my first choice but any animal book would do. In most
of these books, the animals were braver, kinder, smarter
and in general, more likeable than the human characters.
And the people clearly, most of the time, didn't understand animals. They beat them (*Black Beauty*), took them
away from the people they loved and were faithful to
(*Lassie Come Home*), loved and lost them (*The Yearling*).
I hid upstairs in my room on rainy days, curled up under
the covers and wept over Lassie, starved and sick, sitting
outside the school, waiting for her boy. I learned pretty
much every lesson I believe now about being human from
reading about animals.

Ethical considerations didn't really enter into my childhood relationships with animals. Although I did have an
ongoing argument with Wally Johnson, our neighbour,
the local trapper whose job was killing animals. But Wally
was also a wonderfully kind, gentle man who believed
that the only animals that really deserved to live were deer,
trout and songbirds. Everything else he saw as his job to
kill—as many and as often as possible.

We were fascinated by the carcasses of dead animals
in the back of his tiny green Austin pickup. He was always
bringing things to show us. He knew more about animals
than anyone else we knew, and when he sat at the kitchen
table with a glass of dandelion wine, we sat with him and
listened to stories of cougar, lynx, coyotes, beaver, mar-
ten and mink. In those stories, all the animals died. I was

both drawn in and repelled. I didn't mind helping my dad kill the farm animals, but wild animals seemed to me to belong to a different realm, one I sympathized with, even felt akin to.

Wally also knew the woods and mountains in a way that very few people do anymore. When he was in his eighties, he hiked over the Purcell Mountains with a package of salt and fishing line and a twenty-something nephew. With great glee, and while standing on his head on the board swing in the north garden, Wally told me how he wore out the nephew. He had just had a couple of glasses of my mother's dandelion wine. He always did love my mother and her wine.

Wally was always interested in my or my brother's stories of what we had seen in our travels around the farm or in the woods. If we said we had seen a bird or a fish, he demanded to know where we had seen it, what it had looked like, what it was doing. He liked children because he was something of a sad child himself. He had been born in North Dakota in 1900. He often told us stories about how harsh his childhood had been, how little they had to eat and how he had left home at twelve and never gone back. His wife Nettie was the shyest woman around. She wore long skirts and head scarves and made lard-laden, greasy doughnuts which we always politely ate on our visits even though they made us feel sick.

One day we arrived at their house when their truck was stuck in the mud. Wally was sitting in the front seat, gunning the motor and screaming, "Push, Nettie, push," while Nettie struggled along grimly behind the truck, covered in black mud from the spinning tires.

—◦◦◦—

The animals I loved best and thus knew best were horses. I first learned to ride on our neighbour's half-wild horses that they captured, tied in a corral until they were "broken" and then turned them over to us kids to ride. Eventually, after much stubborn begging and pleading, I got a horse of my own. We couldn't really afford a horse. Everything we had we used to survive. A horse was purely a luxury. So my father bought her and then resented her for every mouthful of grass she ate.

He gave me Lady (the name she had come with) for my eighth birthday. My father took me to see her before he bought her. He asked me what I thought. I had no idea what I thought. I only knew she was a horse, a brown horse, a horse that if I said yes, would be my horse. So I said yes.

What he bought, for the enormous sum of $150, was a three-year-old mare whose cunning and intelligence I have never since seen equalled in another horse. When my father hauled her home in the back of the Dodge pickup and turned her loose in the pasture, she headed for the farthest corner and refused to have anything to do with any of us.

What we didn't know, and found out much later, is that she had been "broken" as the term went, by the same general methods the O'Neills used. Someone had put a halter on her head, beaten her half senseless, stuck a bridle and bit on her, ridden her around a little and pronounced her ready to sell.

I followed her around for weeks with bits of apple and oats pilfered from the cows and chickens. I asked my father for advice and he grunted that she was my horse and if I didn't want her, she could go back where she came from. After a while, she let me scratch her neck and shoulders and one day when she was lying in the sun, I lay down with my head on her round warm belly and we dozed together.

Even horses get lonely. One day, Lady walked over to me when I crossed the pasture. She put her head in my arms and sighed the deep sigh horses make when they relax.

Trembling, I went back to the barn and lifted the heavy old leather halter off the nail in the corner above the manger. I went back out. Lady was still standing where I had left her. I lifted the halter up and she stuck her nose in it. I did up the heavy metal buckle.

Then I tied a rope to the ring at the bottom and led her through the gate. I tied her to a tree and went and got a brush. I spent much of the rest of the afternoon brushing her and feeding her apples while she stood with her head down, half asleep. Finally I got the ancient bridle that had come with her and stuck the cold bit in her mouth and wrestled the earpiece over her ears. Then I slid on her back and rode around the house, down the lane, through the orchard and back up to the house. Before I turned her loose, I rubbed my hair all over her hair so I could sleep all night smelling that wonderful salty, sweaty horse smell.

That night at dinner, I announced that I had taken Lady for a ride. No one seemed to think this was remarkable. What to me had been a miracle was passed over between the potatoes and the creamed corn. That night I took my horse-stinky hair to bed and lay all night in a stupor of happiness, dreaming of the places that Lady and I would go.

But Lady soon showed her genius and she got herself, and me, into trouble. She began to figure out how to untie gates, ropes, latches, and open any kind of door. She got into the chicken feed bin and ate herself sick; she tore down a whole line full of clean white sheets that my mother had hung, and trampled them into the mud.

When I woke in the morning, Lady would be tied to a tree in the yard. So I knew she had done something. It was always bad. It always cost something. One night she ruined a whole wagon-load of apples by chewing small bites out of a few apples in each box.

One hot summer night, the Greyhound bus driver knocked on our door. At the last moment, he had seen a black shadow on the road and screeched to a halt. Lady was stretched out on the warm pavement, sound asleep.

I began to tie her in the barn with the door closed. To get out, she had to untie a rope that I had knotted and double knotted, undo the latch on the barn door and then undo the gate. Somehow she managed it. Finally, my father looped a heavy chain over the barnyard gate and wired it shut. That seemed to work.

My father thought that everything on the farm should be put to work. Lady had no job, no use that he could see, except to starve the cows by eating all their grass. Finally he hit on the idea that she might be useful driving the cows on their annual spring trip across the river, where they were left to graze until just before Christmas.

To get the cows to the river, we would drive them five miles south along the highway to the railway bridge. My father went ahead in the truck and my brothers and I ran behind. The cows hated this trip and broke away at every opportunity, into the neighbours' yards or up old logging roads onto the mountains. They stopped traffic and stood stupidly in the middle of the road staring at the cars, while furious drivers honked and waved their arms.

We ran and ran, while our father banged on the door of the truck and yelled instructions. Occasionally, when there was nowhere for the cows to go, we caught a brief breath on the running boards of the truck. When we got to the river, we banged on the rumps of the cows with sticks and rocks until they reluctantly stepped into the river and swam across. Then we could pile in the warm cab of the truck and go home.

Every winter, in order to bring them home, we first had to find them. They were scattered through the brush and marsh and willow thickets at the south end of the lake.

We walked across the railway trestle, stepping over the black, creosoted ties, looking down at the black-green water below. The wind always blew through the trestle girders, viciously trying to snatch us off and throw us into the water. But we made it.

Then it was a three-mile march out to the other end of the dike, to Kootenay Landing, where sternwheeler boats had tied up and received passengers before the railway track was built.

The cows were always hiding somewhere, reluctant to get moving, wary of people after seven or eight months on their own. We had to run through the swamp, through murky black water and mud, leaping from clump to clump of tall salty-smelling sawgrass, while the cows sloshed ahead of us.

After we got them across the river again, we would run behind them all the way home. I once figured out that we had run, almost without pausing, for eight miles.

My father thought Lady should be able to help with

this, but Lady had figured out pretty quickly that she didn't have to do much of anything she didn't want to. One thing she didn't want was to be ordered around by an eight-year-old kid. She let me ride her, but she stumbled and dragged her feet and slouched along and stopped whenever she saw something that might be good to eat. I tried riding her with a stick, the way the O'Neills had taught, but that made her shake and sweat and shy at everything so I fell off as much as I rode her. We weren't much good as cowboys.

My father began turning Lady loose to run with the cows for the summer. I was already riding her less and less. There weren't many places to go really, and there were more and more cars on the road that had now been paved. As well as that, my father now used her as a kind of generic threat. Whatever he thought I had done wrong somehow came around to involving the horse—her expense, her being a nuisance. He called her "a greedy useless waste of time and money."

There was only one way out. One day I said, "Fine, sell her then."

Some people came, a nice enough couple, and then my father told me to ride Lady to Wynndel, where the couple would meet me with a truck.

Wynndel was twelve miles away. We rode there on a June morning, past the wild roses and the swamp full of red and yellow-winged blackbirds, past Sirdar where I went to school, up the long hill and eventually to Wynndel. I slid off Lady's shiny back and handed the reins to the couple. I didn't look back. I slumped into the seat of my father's truck and we rode home in silence. Sometime later I bought a piano with the money, which my mother had made my father save. I bought myself piano books, taught myself to play, began to stay in the house with my mother and tried, almost successfully, not to dream of being wild or of riding horses anymore.

⸺

My sister is now an accomplished rider and trainer. She says that a trained horse with a trained rider can enter into a kind of consciousness where the rider communicates by thinking *do this* or *go there,* and the horse feels the slightest shift in the rider's body and responds.

One night listening to the radio, as I often do when I am lying in bed waiting for sleep, the commentator began talking about a term I had never heard before. Later, I looked it up. The word *umwelt* is a German word that means environment, but it also has a specific meaning in the world of consciousness studies. It was coined back in 1930 by a German biologist named Jakob von Uexküll. Von Uexküll was fed up with the era's dominant view of animals, which considered only how animals acted—their behaviour. He was more interested in what animals experienced, in the texture and quality of their felt, sensory worlds. In an attempt to address this question, he published a monograph called, "A Stroll Through the Worlds of Animals and Men."[2]

To get a glimpse into how animals experience their environment, von Uexküll writes, "We must first blow, in fancy, a soap bubble around each creature to represent its own world, filled with the perceptions which it alone knows."[3]

As we step into each of these bubbles, von Uexküll goes on, "a new world comes into being." Each "new world" von Uexküll called an *umwelt*, "a richly-detailed self-world which corresponds to the unique senses and environments of each animal."[4] By imagining these *umwelt* bubbles, he believed he could also imagine his way into the reality of the animal in question.

My sister does this by thinking and acting, as much as is possible for a predator human, in a way that will make sense to a horse. And she watches the horse for its reaction to her. The horse is also intently watching her actions and reactions. It's a relationship in which they are both fully engaged, and of necessity, highly aware of each other. My sister illustrates just how immediate and sensitive a horse is when she makes the horse step sideways merely by touching her finger on its shoulder.

⸺

At the farm now, I am endlessly conscious and conflicted by the weird ethical contradictions involved in our relationships with animals. For example, many people react by screwing up their faces and saying something like "Eww, ick" when they find out we butcher our own chickens and pigs on the farm. Or my city son-in-law marvels at how, as he puts it, "In the Kootenays, the animals are just as important as the people."

And indeed, at the farm, whenever there is a gathering, we tell endless dog and chicken and coyote and cow and pig stories. After dinner, around the fire, the dogs play with each other or curl up to sleep. All country parties are composed of people attendees and dog attendees. Dogs do love a good party.

So there is people gossip and animal gossip. Both are equally fascinating and equally necessary. The people gossip keeps us informed about our friends and who is doing what; the animal gossip plays a slightly different role. A lot of it is necessary information about how the animals are doing and what needs to be done or not done. In addition, the behaviour of animals is endlessly fascinating and intricate, and we are always trying to understand and come to terms with it.

This year, as we usually do, we bought twenty baby pigs to raise. They came to the farm in the back of my brother's pickup and were unloaded into their new, clean pen. These pigs had never been outside, had been born in concrete pens and raised on concrete. They were terrified to go out, so eventually my brother pushed them out the door of their shed one at a time. And then one of them began sniffing the dirt. And then shovelling through it with his nose. And then tasting dirt and grass roots. Pigs really do caper and kick their legs in the air and this one did. He was manifestly in love with dirt. He kept snuffling through it and then looking at us. If a pig could smile, he did.

The pigs were still in a pen but they had a creek, shade, a mud wallow, grass. Every morning, all twenty baby pigs snorkelled their way through the mud pool. They liked to stand in the mud every morning after I let them out and have an amazingly long pee. The pigs quickly became a tourist attraction. People stopped on the road, brought their children to look, took pictures, wrinkled their noses at the smell and the proliferation of flies and black hornets and asked the same questions over and over. The first one always, "Do they bite?"

At the farm, we love and care for animals. And then we eat them. Recently, we killed five young gorgeous, strutting roosters. But a flock of chickens only thrives well with one rooster. In nature, the others would be driven off and probably eaten by predators. Here, we are the predators, big alpha predators with teeth. This is the bargain we make with our animals—to love and feed and care for them— and then to kill them as quickly and humanely as possible.

These days, attitudes toward animals conflict and clash in every person. My brother walks up the mountain to be with animals. He watches and notices everything but he still rages, much as our father did, about bears in the fruit trees and deer eating the garden. He loves "his" animals but he is still more aligned with our father's values than with mine.

Last week I went to visit a new neighbour who has spent a lot of time and money and energy landscaping his place, making a garden that looks quite natural and beautiful. He has also built a series of ponds on a hill, and each pond is surrounded by an electric fence to keep the otters from eating "his" goldfish. I pointed out mildly that otters are endangered here and goldfish breed so fast they tend to become a nuisance. He shook his head impatiently at me. "The otters live in the swamp," he said as if that somehow justified everything. I decided I liked his garden but didn't much like him.

It never fails to astonish me how much emotion, both positive and negative, people invest in their relationships and ideas about animals. They either love them passionately or, just as often, are terrified of or hate them just as passionately. Stories about animals seem to be either long or short. In either case, they are usually not stories about animals at all, but about people's ideas and involvement, however profound or superficial, with animals.

Which is very odd, because animals don't seem to have similarly passionate feelings toward us—although we matter to them in all kinds of ways. But of course, we don't know because we haven't yet learned to communicate with animals in such a way that that communication matters to us, and most of us still tend to assume we know how they think and what they feel, often without a lot of evidence.

But there is a slow change going on. There are a lot of people working with animals in positive ways, and interesting books about parrots, bonobo monkeys, chimps, bears and wolves have emerged recently—although like most other information not amenable to mainstream thinking, none of this gets covered or talked about widely.

Most of these books are still focused on how much animals are or are not like us: whether they have language, whether they have culture, how they feel about us. But I was very pleased to read about a man named Lynn Rogers, a biologist who has spent time with bears in the northern US woods. Rogers is no sentimentalist. Even after devoting forty years of his life to the black bear of Minnesota, he is under no delusion that his interest is reciprocated. The bears don't really like him, he says.

"June (a female black bear) has no feelings for me," he says. "If she had feelings I think she would want to seek out company like a dog does its master," he says. "But she doesn't think of me in those terms. I'm just the guy that brings her a treat once in a while and that she can ignore and not pay any attention to and that is what makes her so valuable to science."[5]

I also like this quote from a book called *Landscapes of Fear* by Yi-fu Tuan: "We tend to suppress the knowledge that fear is a universal emotion in the animal kingdom from our consciousness, perhaps because we need to preserve 'nature' as an area of innocence to which we can withdraw when discontented with people."[6]

Craig Childs, a biologist who makes a living looking for water in the desert, says, "The life of an animal lies outside of conjecture. It is far beyond the scientific papers and the campfire stories. It is as true as breath. It is as important as the words of children."[7]

Or, as Barbara Noske writes in *Beyond Boundaries*, "Perhaps what I am looking for is an anthropology of animals, a place where the human-animal interface thins and disappears, where 'Otherness' isn't any longer an excuse for objectification and degradation, either in practice or in theory."[8]

The reality of animals will never really be accessible to me or to people in general. But the knowledge of animals is a different thing. People who work with animals or encounter animals on a regular basis—farmers, hunters, animal trainers, etc.—usually have a very specialized and often quite deep knowledge of particular kinds of animals. My sister, for example, although she knows an immense amount about horses, isn't interested in dogs or cats or birds.

And then there's my friend George, a hunter and fisherman who knows his local landscape and the habits of the animals within that landscape amazingly well, but is suspicious and resentful of what he sees as the intrusive meddling of ecologists and wildlife biologists. "I don't like anyone telling me what to believe," he says.

Scientists, while they are extremely knowledgeable about particular kinds of animals, often seem to know little about animals in general. They are constrained by the requirements of science and an almost comic fear of anthropomorphizing animals, which then excludes anecdotal or local or indigenous knowledge. In addition, science seems slow to take up the idea that knowledge of animals gained in a library or through scientific methods is itself biased and skewed to a particular point of view.

Jeremy Narby, in his wonderful but too short book, *Intelligence in Nature*, details cutting-edge research looking at the way both plants and animals think.[9] Much of this research is about what could be termed intelligence, although the term is so laden with human-centric ideas, it makes almost no sense applied to animals and plants.

In fact, one of the difficulties in working with animal "intelligence" is finding a language that works in animal terms. Words like "intelligence" or "culture" don't really make sense when applied to non-humans, but so far we haven't invented anything better. This is a major question for Narby. He explores different notions picked up from scientists he interviews. For example, intelligence is "adaptively variable behavior within the lifetime of the individual," or it's "unconscious information processing," or an "ability to compute and make decisions."[10] He recognizes that all definitions of intelligence come up short, and in the end he opts for a term he picked up from a Japanese biologist: *chi-sei*. It means, loosely, something like "knowingness" or an ability to recognize. It's what Japanese use to translate the English word "intelligence," and it implies a property intrinsic to all nature.

⸺

People who work or live or hunt or depend closely on animals are in relationships with animals that take an almost infinite variety of forms, depending on how such people characterize animals.

And of course, this is the great difficulty—that people are free to characterize animals according to whatever cultural and social framework they happen to be working with: from a woman getting her poodle dyed to match her apartment to the Inuit hunter depending on his dog's sense of smell to get him home. Once, as indigenous hunter-gatherer people, we were all dependent on our understanding of animals. Civilization has been a long process of making ourselves less dependent on animals, on moving away from our connections with them.

~~~

I still spend a lot of time these days with animals, and I am much less certain about what I know about them than I was as a child. But now I listen. And watch. The swallows sit on the porch in the early morning, gabbling and yelling, sounding exactly like a crowd of people at a party or in a restaurant. When the hawk or the golden eagle comes by, the ravens come out to meet him or her. There is obviously lots of communication going on, wing tip to feather lift, and I am deaf to it.

I know something about domestic animals, less about wild animals, almost nothing about insects and lizards and spiders and wasps and flies. I share the farm in June and July with an almost infinite number of mosquitoes, and I truly can't come to any understanding about them because no matter how equitable I am determined to be about our shared life, they drive me quite mad. Screaming mad. Raging mad.

While I am picking raspberries and the mosquitoes are ranging in and out of my ears and eyes, I try to remember we are here together, living our lives in some kind of strange and unknown partnership/relationship, each with our roles and our *umwelt:* mine is heat and berries and itching and satisfaction, and theirs is blood, smell, pursuit, reproduction. In our own ways, we are doing exactly the same things.

But for most people, especially those who rarely encounter animals, the idea of animals remains an area of the unknown where humans can endlessly project their own needs, desires, humanness. In this territory, we lurch from sentimentality to cruelty and back, a lurching that's eerily similar to historical positions previously held by various human oppressors.

It is no longer politically acceptable for men to say what women are feeling, or for white people to assume they know and understand the reality of people of other races. But it is still perfectly acceptable to assume we know what animals are thinking and feeling. But we don't. And can't. And yet these beliefs, whatever they might be, colour and influence all our interactions with animals.

⸺

This spring, a neighbour phoned my house, her voice panicky. The night before, a cougar had broken into someone's chicken shed, she said. The person had surprised it and the cougar had run away. Someone else might have seen the same cougar, she said. Of course they weren't sure, but she was phoning everyone within ten miles with children or grandchildren to warn them.

What I didn't tell her was that my brother had come down from his walk on the mountain a few days earlier and told me he had just found a cougar den with a female cougar and two kittens. We were both glad about it; there are too many deer and not enough predators in our neighbourhood.

⸺

Neither of these stories is a judgment. One person is terrified of cougars and one is not. My brother walks up the mountain every day as he has all his life. He walks up to deer, ravens sit on his shoulder. He's not a Thoreau kind of guy. He's a redneck logger who loves the place where he lives and knows enough about it to walk through it with a sense of comradeship rather than fear.

But my neighbour's fear is a lot like being terrified of terrorists. If they never attack, the unfearful people can crow triumphantly (after a long while) that nothing was ever wrong, but it only takes one bear/cougar/wolf/coyote attack for all kinds of stories and fear to circulate and for the fearful people to consider their fear justified.

Whether any of the stories of people being in danger or hurt are true or not, whatever caused the animals to act the way they did never seems to be the issue.

As Nigel Rothfels wrote in 2002 in the introduction to his anthology, *Representing Animals,* "The way we represent animals is deeply connected to our cultural environment and this cultural environment is rooted in a history... who controls that representation and to what end it will be used will be of profound importance in the coming years."[11]

We can never actually know what animals feel or think. What we believe, or choose to believe, about animals and the ways we choose to interact or live with them have political, environmental, social and ecological consequences. We see our fellow species on this planet through a haze of romanticized, sentimental or cruel illusions. The philosophical, cultural and historical assumptions about "nature" made by scientists such as Descartes and Newton have no basis in science whatsoever. Such assumptions justify cruelty, exploitation and sentimentality, but they do nothing at all for true understanding or communication. That will have to come from some radical new thinking and ways of interbeing with our fellow species.

Chinese Swallows

In April, I drove around the Kootenays, going to libraries and reading to school kids. I got back from all the schools and libraries and kids with shiny faces. I cleaned and slept and then got in my car and drove to Vancouver to be the nanny while my daughter ran a film festival. And then after a week or so in car-choked Vancouver, I got back in my own car and drove home again, arriving somewhat bewildered and exhausted and with piles of stuff to do on my desk, in the house, in the basement, in the garden. Laundry, letters, bills, edits, emails. Then finally the long weekend came, the sun shone and I went forth to plant the garden.

Out I went, and in the meantime the Alberta summer cottagers next door had arrived for the long Easter weekend. They unloaded their lawnmowers, weed whackers, chainsaws, leaf blowers, power washers and vacuum cleaners. The highway howled with RVs, motorcycles and people in too-fast cars passing each other on blind corners. After I weeded the garden, I sat on my deck in the shade in the rocking chair and, despite the noise, fell into the deepest, soundest and most satisfying nap ever until the swallows woke me. I was still half asleep, and as I listened to the swallows yelling away over my head, I heard the tones and shifts and thought, No wonder I can't understand them. They are Chinese swallows.

Next Saturday is a wedding dinner on the lawn, picnic tables and flowers—turkey, lamb, potato salad, and I am *ordering, not making, the cake. I will go to Cranbrook with my friend Yvette and buy wineglasses and red tablecloths. Such is life in the fast-paced Kootenays.*

Machine Noise

EVEN WHEN I DON'T SEE him, I can tell when my brother has arrived by the *pup, pup, pup* of some machine starting. Machines and their various noises punctuate my days. I'm always pleased to hear him at work, although I think privately that he and my father never saw a job that they didn't think could be done better with a machine. I struggle along doing stuff by hand: digging, weeding, planting, pruning and picking. I'm not a machine kind of person and it shows. I can drive a car but if there were an alternative involving horses or walking that let me buy groceries and return home in a reasonable amount of time, I'd happily take it.

When my grandfather died in 1967 and left my parents a small legacy, my father did what he had long wanted: he stepped out and bought a bright yellow backhoe. As soon as he hauled it home in his old orange and blue dump truck, he backed it off the truck and began digging up rocks. From then on, early farm mornings were punctuated by the sound of the backhoe warming up. Over the next few years, my father reshaped the topography of the farm. He built roads, fences, a pond; he planted trees, harvested firewood and logs and even built a log house.

When I was a child growing up on the farm, a background of constant sound, both natural and machine made, was a part of farm life that I took for granted. I was an adult when I began to realize how much sound mattered and how it affects the inhabitants of a place. Noises on a farm are constant, muted and important. People often remark on how quiet the farm is but, in fact, it is only really quiet late at night and even then there is usually the rumble of wind in distant trees, the lake shushing in its rocky bed, crickets, frogs, night birds.

Sound is also seasonal for both animals and machines. The ubiquitous fall/winter sound is the chainsaw, signalling another wood supply in the making.

In the winter, great horned owls call from the mountain, and the voices of winter birds—flickers, jays, chickadees—haunt the bushes. Early spring is signalled by the dawn chorus of swallows and robins, and the late evening chorus of frogs. In summer, the night is full of the sound of crickets, nightjars, cicadas. Perhaps the only really quiet time is late fall, when everything pauses before winter except for the occasional machine noise accentuating the quiet.

Layered over these noises are the farm noises: chickens, pigs, cows. On the farm, the absence of noise is usually a good thing; a cow or a chicken making an unusual noise is a sign something is wrong. Over this are the noises of the many, many machines that are a reality of modern farm life, a reality of our age.

I always have conflicted reactions to machine noise. The noise of machines working on the farm where I live doesn't bother me. In fact I find it exciting and anticipatory. It can be the Rototiller preparing the garden for planting, the chainsaw bucking wood for winter, the backhoe digging a ditch or a pond or holes for new fruit trees. Machine noise on our farm, to me, means work, food, production. I know what it is about and thus I can bear it. But I also know that the noise of machines is often the noise of destruction. It usually means something non-human is either being intruded on or destroyed. Someone's—usually a non-human someone's—home is being invaded once again by humanity.

My father's father farmed the Saskatchewan prairie with horses. My father, like many men of his time, and out of necessity, was a genius in understanding, fixing and maintaining machines. The backhoe was the only machine he ever bought new. Everything else was second-hand, and like many other rural farmers, he kept a junk pile from which he could raid parts that even when they weren't quite right, could be made to fit. He spent a lot of time both fixing and cursing his collection of half-dead, beat-up machines. It was another war in the many wars against the non-human world that made up his life. But he usually won and the machine eventually rattled into furious life.

Every year I go away for a few days to a nearby camp on Kootenay Lake that has no electricity. The camp is on the edge of an unsettled peninsula in the lake. There is the occasional and very subdued rumble of truck noise from the highway across the lake; a few boats either putter or roar by, depending on whether their inhabitants are fishing or playing. But the rest of the time there is no machine noise: no background hum of electricity, no clanking, banging, buzzing, roaring or ticking. It's always a shock to leave this place, even after a couple of days, because most of the time, along with many other people, I successfully manage to block out a lot of the background machine noise I live with. But of course it is always there and on some bodily level, I am always aware of it. I can only realize this in its absence.

———

Right now, there are many people near the farm building houses. Building a house is an amazingly noisy affair; almost nothing about it is quiet. A house initially needs a road, power lines, water lines, a sewage system, all built with machines, and most of all, a house has people and cars. Every house seems to concentrate noise and disruption within the broader landscape. Even building the smallest and most inoffensive house requires a lot of machines: backhoes, cement mixers, saws, nail guns and hoist trucks. A house is itself a kind of machine where people are fed and groomed and kept safe by the walls and systems of the house.

Lately, when I go outside my house, I am up against a wall of noise. Bulldozers, chainsaws, skill saws, cement mixers, the machine-gun splat of automatic hammers.

I find myself making excuses to not go outside and then fantasizing about burning down their houses, smashing their windows, putting roofing nails on their driveways. I will do none of these things. I try, unsuccessfully, to be patient, to tell myself these people are making homes that they will also love, that the noise will end. The people will come for brief vacations, and then, I hope very much, they will go and take their noise with them.

———

It isn't much, as noise goes. I used to stand on a busy corner in Vancouver almost every day, changing buses either to go out to the university or to come home again. The Broadway–Cambie intersection is a constant barrage of trucks, buses and cars. I always braced myself for the wall of noise, so thick I felt I could lean against it.

I also learned to hate the Cambie bus. It was always late. It would come groaning up the hill, full to the brim, creak to a stop, while the line of us stood bareheaded (having put our umbrellas down) in the pelting rain waiting to see if enough people would get off, so that we could get on and get home. Sometimes the bus driver would wait until everyone had crowded far enough to the back to fit us all on. Of course once I got on, the next problem was shoving and squeezing my way through the damp mass of bodies to get off again. It was always such an amazing journey to get home and shut the door, against the rain, the cold and most of all, the noise.

It's all very odd. Of course I have machines. When I lie down in my house at night, the fridge gurgles and hums, the clock ticks, the freezer in the basement has its own vibratory tone that I can't so much hear as feel. The radio hums away quietly all day while I work.

There's an odd conflation of machines with progress. Somehow having machines and making a lot of noise with them is seen by many people as being on the side, if there can even be said to be sides, of civilization and progress; and somehow, civilization and progress also tend to be conflated with good things like medicine and education. This weird blend often shows up, for example, in online discussions of things like going back to farming without machines, a move that is sometimes seen by responders as sending us all collectively back to the Middle Ages, complete with dirt, fleas and disease.

At some point in my childhood, we farmed with horses, and we used kerosene lanterns for light. We had a reasonably comfortable existence, and so I don't worry too much about whether the machines will all one day stop— whether from energy shortages, or some natural disaster, or simply from utter complexity. In fact, like many people do, I want them to stop. I long for them to stop or at least slow down.

In the morning these days, I sit quietly with my tea and read various pundits on the internet, who predict that the whole world is engaged in a transition that will take a long time, that will result in humans living without the same number of machines because eventually—and probably much sooner than we expect—we are going to run out of affordable oil. And then the long, drunken noisy machine age will change into a different mode. No one presently has any clear vision of how this future world might function—although there are endless arguments about it—but then no one in the eighteenth century could have possibly envisioned the world we live in now.

Envisioning the future has generally been the province of futurists and science fiction writers, who have usually been wrong. I try to follow the complicated arguments that rage back and forth between geophysicists, economists and other experts. To most people, a future without machines seems almost unthinkable, and many of the arguments tend to focus on just how much time the world has left to prepare for such an eventuality, and whether we can create enough alternatives to oil in order to continue our present technological madness into some blurry future time.

But time is an odd construct and the future is invisible. At the farm, there are many kinds of demanding time. I carry a kind of farm-clock in my head. I snap awake at 6 a.m. I have a schedule: the usual daily working time, time to write, to garden, to stop for tea, to do chores, but none of this schedule has much to do with a clock and everything to do with what needs to be done.

And outside this schedule, the timeless time of mountains, trees, rocks, plants and soil that I try to enter every day. I have moments sitting on the deck in the blue-green late afternoon, or at the beach on a golden-hot August day when I feel this happen. Time lets go. This is not enlightenment, but it is a relief.

Although I have tried, I have never achieved this in a city or around machines or when the distant sound of other people's machines cuts the air. At the farm, the sound of machines smashes eternity into bits. The distant noise of machines drills into my head and heart and snaps me to attention, back to now, back to irritation, and I have yet, despite some effort, figured how to shut it out, or change its effect on me.

———

Machine noise is about work and striving and restlessness. It is about change and those peculiarly loaded words, growth and progress. Only people have machines and only people use them, loading the atmosphere with noise and restlessness. It is those moments of racketing intensity: boats flashing over the serene summer lake, chainsaws bringing down rank after rank of trees. It's happening all over the world, the noise, the racket, the banging up against, the crashing—through rather than into—that feels like a war. Which it is.

There is a man in the United States who has gone all over the country looking for one square inch of land where there is no human noise. He thought he had found it in the rainforest of the Olympic peninsula, but then discovered that several times a day, jets flew over. He wrote to the airlines to see if they would move their flight paths but they declined. There is no height, apparently, to which an airplane can ascend where its noise will not be heard on Earth.

Lots of information is known about the health-destroying, sleep-destroying effects of noise. Less, it seems to me, is known about why people love noise. They must love it to produce so much of it. For example, people in the Okanagan Valley in the summer are subjected to the thunderous noise of jet boats. Because the valley is bowl-shaped, the noise echoes across the lake and into everyone's houses. And also every summer, pack after pack of motorcycles roar by the farm. Our twisty narrow mountain highway is advertised in some magazine as a great challenge for motorcyclists.

No matter where I am or what I am doing, my head snaps around when they shriek by. No doubt there is some delicious and thrilling attraction to being part of this speed and noise. I imagine it as being the direct opposite of my bemused trance on the porch—instead of intrusion, it's a song of joy, a song of adrenaline and power, a song that blocks all thought and forces the rider into another kind of timelessness, one that only lasts from brief flashing moment to moment. Maybe that is a relief of some kind as well.

But it's not a relief to me.

It's not necessary for jet boats or Harley-Davidson motorcycles to make as much noise as they do. They, in fact, can have mufflers. But owners take them off. The noise, for the riders, makes a statement of some kind, which I'm not privy to, because I don't and couldn't own one of those machines, and I don't understand why someone riding a Harley wants me to hear them. What do they think I'm hearing? That they exist, that they're present, that they matter? Which isn't, of course, what I'm hearing at all. What I hear is the moral and physical equivalent of the middle finger, writ large in the broken churning air for minutes after they have passed by.

Every noise carries a message and makes meaning. Machine noise, however meaningless it initially appears, carries a load of political, cultural, social and environmental messages. Most of them we manage to ignore most of the time, or we couldn't live in this world we have all created together. Machine noise, however benign its intentions, still shouts about intrusion, invasion.

I go to bed early at the farm. I lie in the dark, drifting, listening to music or CBC radio. One night, I heard a discussion of what the world of whales and dolphins might be like. According to the researcher, it is a world made primarily of sound, of vibrations that can actually penetrate inside other beings rather like ultrasound, so that it is possible that whales can hear and understand each other's feelings. Whale songs can be heard for astonishing distances. A blue whale song off the coast of Newfoundland can be picked up by a microphone in Jamaica, an ocean basin away. Humpback whale songs are complex symphonies sung in unison by many male whales, constantly evolving over time. I think about sound as I fall asleep in the wonderful blanket of silence of the farm at night. I wonder what the whales hear, in their great blue-black depths, as the freighters and submarines and oil tankers churn by. And do they sing about it?

If the machines fall silent one day and the motorcycles stop roaring by and the airplanes stop splitting the sky in two, what will the people who love their machines and their noise do? How then will they find a way to lift a contemptuous finger to eternity?

Lush

The garden is in yet again. The ground is weeded. Three weeks of rain soaked everything, and the fir and cedar have responded with lush electric green tips on their branches. The hot-weather plants, tomatoes and eggplants, are sulking and yellow but their turn will come now the sun is back. Right now they are pushing their root tips into the soggy fertile ground and getting ready to produce mountains of fruit. I planted the large, purple-black eggplants again. Every year I pick them and stare at them. How can a purple be so black, a black be so iridescent? It is impossible to leave the farm at this time of year. There's always something that needs doing, something to plant, something to weed, something to water, even something to pick. Even though it is only the end of May, the garden is bursting with spinach, radishes, lettuce, onions, Chinese cabbage, Swiss chard, and even broccoli. Every morning, I wander with my coffee. There is always a new flower opening.

And often, late at night, unable to sleep, I read. I read about the oil plumes in the Gulf of Mexico, albatross chicks starving to death after their parents feed them plastic, mistaking it for fish. I read about peak oil and possible food shortages in the future, about global warming, ice sheets melting in Greenland. This fall, I will fill the shelves again with canning, jars full of dried fruit, and a freezer full of vegetables, fruit and meat. Life continues.

And yet I am planning on running away from it all again. Not for long and not soon, but I need to finish the two books I have been slowly working on. So I will try to spend some time in the city in the fall. The farm is a difficult place for a writer. A farm needs to be a community; it needs people; it needs parties and dinners and planning and work. And I need solitude and time to walk and think and write. So I will run away again to the smelly city where the grocery store is full of expensive fruit that I would never pay for at home, but the library is just down the hill and all the tools I need and want as a writer are there: my friends, books, my writer's life. And all the time my heart will be crying, go home, go home, go home.

And so I will leave again and return as I have so many times.

Today the clouds are rolling in, but the tomatoes are in flower as are the intensely blue Chinese delphiniums, the purple delphiniums, the white miniature roses and the pink poppies that are ready to "pop."

Whenever I go for a walk, I feel like cheering for them all.

Houses

WHEN I WAS TEN, MY father began building a new house. He drilled holes in the bedrock of the granite hilltop behind the existing farmhouse with his ancient compressor and blew out an almost square shape that would become the basement. We children helped him stuff the sticks of dynamite in the holes and then pile rocks and boards on top. We thought nothing was more fun than blowing things up. Some of the dynamite sticks were so old, they were damp with nitroglycerine that had oozed through the paper.

"Don't shake your fingers," said our dad, half joking, half serious. When a shot of dynamite was ready, and after he'd lit the fuse, he'd yell at us, "Get down and open your mouth." That was so our eardrums wouldn't blow out. Then the shot would go off and rocks would rain down all over the farm. Because of the old dynamite combined with stumping powder made from fertilizer, it was hard to regulate the power of the shots.

Over the next twenty years, my father worked intermittently on the house. He cut trees for the log walls. The trees had to sit for a year or two to dry. Then he hauled them down the road he had made on the mountain—the logs piled on his homemade trailer, which he towed behind the ancient Farmall tractor with no brakes, held back by the farting, snorting motor.

Slowly, log by log, the walls went up. Then he cut more trees, bought a small sawmill and made all the lumber for the house, board by board, then planed the boards on a small planer. He worked at night with a trouble light and a chainsaw. He worked on weekends and in the winter after the farm work was done. I didn't see a lot of the building. None of us children or even my mother paid much attention. In fact, most of us forgot about the house. And then one day, my mother picked up her favourite saucepan and walked across the yard and moved into her new house, the house with a view.

When pioneers built a house, their concern was to keep it away from the wind and damp. So the original home-steader on our farm built his house in a hollow beside the road, with no view. That was the house where I grew up. It now sits empty, and I live in this log house that my father built for my mother, on a hill, with a view.

The room upstairs in the old house where I played and read and studied and slept echoes when I walk into it. When I was eight, my mother gave me cans of house paint and encouraged me to paint horse murals on the walls. They are still there, hidden now under torn, purple-flowered wallpaper.

The room faces north. Out of the window, I once stared into the green depths of the walnut tree, from which squirrels would come and perch on my windowsill to scold me. I read the first book I ever read in that room, sounding it out, word by word, and after that I read and read—on rainy afternoons, devouring books and apples— or hiding under the blankets at night with a flashlight—or when I was supposed to making my bed: book after book, story after story, dreaming myself into becoming a writer.

The room was my refuge from the rest of the family. I had a desk in the corner by the chimney and from there I could hear most of what was going on: my mother's footsteps as she went back and forth from the stove to the fridge, the slam of the back door as my father went in and out, the sound of their voices arguing about money.

In the spring, the swallows in their nests under the eaves woke me at first light, and as I woke up, the bats flew back from their night's foraging and crawled, squeaking, into their tight spaces under the rolled roofing. Outside the deep and familiar silence of the farm was broken by the cows lowing for grain and the sound of the milk pail swinging from his hand as my father walked to the barnyard.

My brothers and sister and I believed our house was haunted. The man who had originally built it, a Frenchman named Pierre Longueval, had drowned in the lake. The house was a spooky place. The rooms were small and dark and each had a door. When we first moved there, we didn't have electricity, and the kerosene lamps made dark shadows in the corners. Our mother had a talent for telling ghost stories, and our father thought it was funny to jump out from behind a door when we were heading down the hall for bed.

After my parents moved into the new house, I lived alone in the old house for ten years. I was over forty. My children were grown. I had quit teaching and was determined to succeed as a writer. At first, the house was spooky and dark. My cat hissed and arched its back at shadows. Things clanked and banged. One night, finally, with no real hope or belief at all, I walked through the house with burning sage and cedar and said, "Be at peace."

The house—or I—calmed and then gradually it wrapped itself around me and became my home. Even now, when I walk in the door, abandoned and empty as it is, it still feels, smells and sounds like home.

This spring, my brother tore the insides out of the old house, the first step towards renovation. I walked through the house, still my house, but the sense of it fainter now, almost not a house, almost just a hole in the air—open windows, cracks in the roof. We burned wallpaper and boards and old pee-stained mattresses that our father had stuffed in the attic for insulation. Pierre's ghost drifted away in the smoke. When the house is finished, in some unknown future, I hope that once again it and I can live together in peace and tranquility. I go visit it. I talk to it. I whisper, *I'll come back*. Whether the house believes me, and whether I even believe myself, is still unclear.

The log house where I live now is the farm's house, and therefore all kinds of people wander in and out on a daily basis: family members, my friends, visitors and various dogs. Many days, the teapot fills and empties and is filled again. I feel it behooves me to keep the place at least minimally tidy and in working order.

But I have never been a house person. As a child, for me to be given a choice between being outdoors or in was no choice at all. I was almost always outside. The house was a place to eat and sleep and read. My mother looked after the house. Talk of housework, or of decorating or moving furniture was an immediate excuse to flee.

But now I have this house and it always seems to want things, drapes and flooring and cleaning and furniture. Occasionally, on my way home from buying groceries and chicken feed, I visit the local hardware store and wander, lost and marvelling, up and down the aisles. What a lot of stuff one can buy for a house. I stand in front of the appliances, coffee makers, blenders and food processors. I can't believe I am doing this. I hate stuff. I am an anti-consumer. But there I am, staring at stuff. The house is making me do it.

The stuff that my house is full of goes unnoticed until I try to do without. Louis, my grandson, comes to the farm for the summer. Lately, he has been talking about a TV show he watches, in which a group of kids are attempting to survive in the wilderness. We have decided we will play survivor on the beach this summer. At night, as he goes to sleep, we have conversations about what we can take with us. Are sleeping bags okay? Yes, he decides. Can I take my Swiss Army knife? A pot, teabags, salt and butter? I know where this leads. I've packed up many a picnic for the beach. The amount of stuff needed to produce just one meal is formidable.

When I was a kid, I loved the idea of surviving outside. I had various hideouts on the mountain above the farm. Often, especially in the spring or fall, I would take some matches, a can of beans out of the pantry and my trusty hatchet, and head up there just for the delight of making lunch on my own. There were two books that I loved and read over and over: *Two Little Savages*, by Ernest Thompson Seton, and *My Side of the Mountain*, by Julie Craighead George. Both are about kids who lived in the woods and did it well. I never quite lived outside, but I always liked to think that I could, if I had to.

But at sixty, a teacher and writer, I find myself stuck in the house far more than I would like. And I am completely amazed to find that I am actually learning to care for this house, learning how much time and energy and stuff it takes to keep a modern house functioning. And this isn't a big middle-class house. This is a small log house on a farm, with only a woodstove for heat. And outside—lawn, garden, mowing, pruning trimming. Caring for a house seems to be endless and I resent it and go on doing it at the same time.

My relationship with houses has always been a puzzling one. Despite my new interest in caring for this house, my kids tell me I am only camping in the middle of it. It's true. Houses are mysterious to me.

My physical and emotional life is lived intensely out-doors where the décor comes ready-made and needs no improving by me. My intellectual life is lived indoors but isn't dependent on my surroundings. In fact, when I am writing, I am pretty much oblivious to my surroundings and can write just as happily in a motel room as at my desk.

But I see houses and people's dreams of houses, the refuges they make of them, the energy with which they imbue their walls and floors and windows, as if perfecting an exterior manifestation of their lives will protect and smooth over their inner life and give it meaning. And of course, the North American consumerist capitalist culture loves this. The media and magazines collude in this, urging people to spend money to collect furnishings, to plan colours, to decorate, to buy new appliances and other additions to their houses. People's relationships with houses have changed over the last thirty years. A house was once a shelter, a home, a place to raise a family. Then it became something else—status symbol, palace, place to show off.

Perhaps those who love their houses think that once everything is just right, life will freeze itself into a model of how they can be in this world. But it never seems to work that way. There is nothing sadder than a house full of cherished stuff when the person who cherished it dies. The only thing that gives a house a life is the person who lives there, whose life seeps into the walls and windows, the house serving as a second coat, sometimes a suit of armour, sometimes a disguise.

During the seventies, our part of the world was full of people who had moved from the city in search of "the land." People who had never built anything more complicated than a birdhouse in shop class. Now they were all building houses. Conversation at parties among the men was full of discussion of concrete, septic tanks, skill saws and mitre joints. The women clustered in whatever room was serving as a kitchen in whatever half-finished house we were in, talking about the difficulties of raising children without running water.

Most people did similar things. They built part of a house and then moved in with the theory that they would finish it while they were living in it. They rarely did. Everyone lived with pink insulation, plastic walls, plastic-covered windows and plywood floors.

One of these houses was built by my best friend and childhood buddy, Alan. A few years ago, he died very quickly of brain cancer. His children were grown and after he died, his wife remarried and moved away. The house has stood empty for about seven years. It has taken on that gaunt and unloved look that empty houses all have. Partly caused by decay, unmowed lawns, unclipped bushes, unraked leaves, but not just that. I am not particularly superstitious about such things, but this house wears an air of menace. I took a friend there once, and she was immediately stung by a wasp.

I tried to go there for coffee with Alan's widow not long after he died. When Alan was alive, I went there two or three mornings a week for coffee. But this time, when Joanna brought her coffee machine into the house and filled it with coffee and water, it spewed water, steam and coffee grounds all over the floor. Twice. Her dog had grown up in that house. My dog was used to coming in with me for treats of biscuits and bits of hamburger. Neither of them would come in the house despite our coaxing. We gave up and left. I haven't been back since.

～

A house is a story, and the life and death of houses is akin to the life and death of stories. Like many writers, I am superstitious about stories. I know they come alive; they live in my mind and while I am writing a story, it churns like a bright ball in some corner of my mind. But a story can also die.

Once I had a novel that was almost dead and it was a particular house that helped save it. I had been moving and changing my life and trying to write at the same time. I had just re-entered university at fifty. I had moved away from the farm and moved in with my grown daughter. I had left my aging mother and father behind, and I knew my mother was worried about her memory and so was I. I was terrified that she would die while I was gone, or that my father would sell the farm, or that my old dog would die.

And then while I was at university, I got an invitation to go to Dawson City to stay in Pierre Berton's childhood home and write. So I did.

Pierre Berton's house is a small white frame building. Many writers have now stayed there. It has been added onto and modified since Pierre was a boy. I was there from September to December. It had an oil furnace that came on at night with a roar like a jet engine taking off. In early October, I got a call from someone at city hall telling me to turn the tap on in the furnace room and leave it running. That's how the city kept the water from freezing in the winter. The water pipes running through the permafrost were heated.

Every morning and evening I wrestled with my almost-dead novel and tried to breathe life into it. In the afternoons, I walked, went to the coffee shop or the doughnut shop, or I stared at the broad grey racing Yukon River and came home to tea and the brief afternoon twilight.

Jack London's cabin was just across the street, small, closed off and uninviting. I didn't love the north. I didn't love the cold. I sat at night and read books about the gold rush, about men huddled in tents through the long dark who hadn't known enough to bring candles. I finished the novel.

—∞—

I have often said and it is true, that I belong to this place where I live far more than it belongs to me. But belonging to a place, versus owning a place—or being owned by a place versus being possessed by my sense of ownership—are very, very different ideas and states of being. I don't want to own this place, but more than that, I don't want my sense of connection to be transformed into one where the place, especially the house, somehow owns me.

What I want is for this house and I to live here in peace. As I plant fruit trees, shade trees, berry bushes, gardens, the house has begun to create its own ecosystem. My brother built a pond just below the house; he built this pond because he could (he had the machines), because he wanted to, and because we both know that any pond or puddle of water will attract wildlife. So now barn swallows and cliff swallows live in colonies of mud nests clustered together on the roof above my deck. They eat mosquitoes and are sheltered from hawks and predators by me and my house. They swoop over the pond and eat bugs. Among their mud nests are many small cones made by paper wasps. Why they live with swallows I am not sure, but their presence seems to keep the much fiercer and more predatory yellowjackets away. As I showed my grandsons, you can stick your nose right up to a paper wasp nest without disturbing them. The coyotes hunt mice all winter in the long grass around the pond below my house, which disturbs the dogs but not me.

But I am perplexed by my perception of houses and the contradictions in what I feel about them. I hate the intrusion of other people's houses onto the landscape and yet I love my own warm den. When I come home on a wet, dark November evening from the long drive to town for groceries and dog food, there is nowhere else I want to be except in my house, in my chair, under a blanket, solitary and comfortable and not moving.

And yes, I recognize that any house is a huge disturbance in the natural world. A house needs to be connected in a myriad of ways; a house means a road, a power line, a water line, sewage; it's a box that sucks in energy and water, that displaces everything that used to live where it now sits; a box that to exist requires that dirt be moved and materials be brought from far away; a box that eats resources; a box that destroys all of the natural world that was there before it arrived. Over the past few years, the summer houses along the lake have gotten huge, elaborately, ludicrously so. I can never imagine the minds of the people who live in these houses and yet, on the few occasions when I meet them, they seem perfectly ordinary people who happen to live in enormous houses. My heart sinks a bit every time one of these houses goes up. People build closer and closer together and each new house means that an area once "wild" is now more civilized.

A tract of houses covering mile after mile of the earth's surface with the trees removed, the watercourses buried and the land covered with mowed grass is not even an ecosystem. Calling it a desert is too kind. A desert does have an ecosystem. A suburb is something never seen in nature at all, a machine of sorts that sucks in energy, goods, food, plastic, electricity and gasoline and returns nothing. It succeeds brilliantly only in letting its inhabitants believe they have ensconced themselves in some sheltered bubble where the reality of the natural, living breathing system all around them can never enter in. It's like living in a human fairytale that will slowly and inexorably become a nightmare.

I am comforted to know that at least the houses at our farm, both the new and the old, have become boxes of stories. These houses, too, have lives of a sort, even if it is only in the stories we tell about them and the stories that have been generated by the lives lived within their shelter.

Landing

I awake one morning at 4 a.m. as the birds begin their dawn chorus, realizing I have finally, truly landed after three weeks away working and teaching. I get up and drink my tea on the deck and watch the light come across Kootenay Lake.

In the afternoon, it's blazing hot and I go to the lake. When I wade into the water, my thought is that it was worth waiting a year, a winter, a cold spring and three weeks in the city just for this. I swim and wash my hair and sit under a tattered umbrella and finish my cheap mystery paperback and stare at the light on the water. Then, back in the house, I wash sheets and make the bed and sweep the cat hair off the floor. For the rest of the summer, the house will host many visitors. In August, I'll host another writing retreat.

The farm was busy when I got home. My brother and his son were pitchforking hay into a pile. My sister-in-law was picking raspberries, the cherries were ripe, Saskatoons dripping off their bushes. Lots of garden stuff as well: peas, beans, red cabbage, beets, onions, garlic. It all needs pulling or picking.

Teaching at the UBC Creative Writing early summer session was fun. I had a roomful of very bright students. I taught in the afternoons so mornings were for editing and reading excerpts for discussion, and then late afternoons, there were panels, dinners with students and then often an evening activity as well.

So finally, summer can really start. I want to catch every moment of it. It's so short and winter is so long. Back to the beach I go for long sessions under the umbrella, beside the summer table laden with books, my drawing pad, pencils, a couple of peach ciders and maybe some fresh cherries.

Writing Land into Language

> Where does that mystery lie? Is it in the fact that place has a more lasting identity than we have and we unswervingly tend to attach ourselves to identity. Might the magic lie partly, too, in the name of the place, since that is what "we" gave it? Surely, once we have it named, we have put a kind of poetic claim on its existence.
>
> Eudora Welty, *Place in Fiction*[1]

THE SUMMER MY GRANDSON LOUIS turned eight, his mother and father let him come from Vancouver and stay at the farm with me without them hovering. After he settled in, one night after dinner, he looked curiously at the log house.

"Who built this house?"

"Your great-grandpa."

"All by himself?"

"All by himself. He blew up the rock for the basement, cut down the trees, made the lumber on his sawmill, and built this house. It took about twenty years."

We went down the stairs to the basement and looked at the places where Great-Grandpa had drilled holes in the rocks and blown the basement out of solid granite so he could build the house on top of it.

"Can I blow something up?" Louis asked.

"Sure," I said. "Ask your great-uncle Bill. He loves to blow things up."

As we went back up the stairs Louis said with great satisfaction, "Great-Grandpa was a real man. A man among men." Where he might have heard such an antique phrase escaped me. He said it several times with great satisfaction and I could see it went someplace in him, solidified his sense of who he is, and who his family is, and what can be done when someone is a "real man."

⸺

A while ago, I went, feeling skeptical, to a university conference on literature and the environment. One panel was a group of academics talking about farming. I sat there becoming more and more irritated because it quickly became clear that none of them knew anything about farming. Near the end, one panelist who was discussing farmers in Taiwan talked admiringly about people there being "married to the land." A woman in the back of the room reacted in horror to the idea.

"Sounds like slavery to me," she snapped, or words to that effect. Her face registered her alarm and resistance to the idea of being "married" to land and somehow, therefore, tied and trapped, as if to some intransigent and implacable spouse. As she hurried from the room, I wondered what I could say to her, I who am happily bound to one particular place.

Days later, my friend and I drove home from the conference. Several hours into the trip, my bum hurt, my eyes burned, and driving started to feel dangerously automatic. But when I made it to the Kootenay Lake ferry, I began to feel, as always, at home, in my community, in my neighbourhood. By then it was late, dark and cold. During the fifty-kilometre drive to the farm along the twisting road, watching for deer and elk, I squinted my aging eyes into the (fortunately) rare oncoming headlights, and dreamed of my log house perched on its granite bed above the lake. My bed, I thought, longingly. My coffee pot. My dogs. My chair. My radio. Silence. Solitude, with dogs, cat, books and email.

In the morning, I woke to the familiarity of it all. It was June so I took my coffee and went out. I ambled, examined each bush, tree, rock, blade of grass, new flower and plant—what's up, what's new, what's my brother been up to. I noticed what needed doing, what had been done, what the next gardening task would be. I greeted each ancient piece of machinery, examined the trees, the budding fruit, the blossoming raspberries. I re-connected, re-rooted myself, all my nerve fibres, all my being, back to this land.

All my life, my existence has been bounded by a strong sense of home. When I am away, I miss it. I am nostalgic for it. Once this sense of missing home was written about constantly, particularly in songs. Think about all those sad, yearning "folk" songs, songs about leaving a homeland, songs about loyalty to the place where loved ones are buried.

For me, the idea of not living in my home is inconceivable. I am where I live. The place has formed me and bound me. If I were to be forced to leave my home, I would indeed be "home/sick," nostalgic. I would no longer know how to live.

How the word "nostalgia" has changed. It used to mean, literally, "homesickness," *nostos* being Greek for "return home" and *algia* from *algos* meaning "pain or grief," and it was considered a potentially dangerous illness even into the 1940s. As William Fiennes notes in *The Snow Geese*, "in the late eighteenth and early nineteenth centuries nostalgia was widely reported in the European armies, especially among soldiers who had been conscripted or impressed... Weir Mitchell, a doctor in the American Civil War reported that 'cases of nostalgia, homesickness were serious additions to the peril of wounds and disease.'"[2] In a 1943 edition of the journal *War Medicine*, Fiennes discovers a definition of nostalgia taken from the *New International Encyclopedia* published in 1905: "Nostalgia represents a combination of psychic disturbances and must be regarded as a disease. It can lead to melancholia and even death. It is more apt to affect persons whose absence from home is forced rather than voluntary."[3]

Generally, this meaning of nostalgia as a sickness resulting from a longing to return home has been lost, or perhaps erased; now it is more often seen as a looking back in time, more than a looking back to a place, and it's often seen as a kind of syrupy sentimentality.

───── ∞ ─────

The relationship between a person in his or her place, where she or he is known and knows what it means to be there, can be as intense and powerful as that between the lover and the beloved. I know this land as I would a lover, my body, my friend.

And of course, there is a difference as well.

Land is indifferent to us, or so we believe. The eternity of land both belies and seduces all our feelings about it. No matter how much ego you expend on land, how much you dig and build and blow things up, the land remains unmoved, inexpressive. So loving the place where you live isn't a romance at all. It's something much older and deeper. Not about patriotism, or identity, but about home, being in a familiar ecosystem, knowing and being known. It's about biology and survival and biophilia and what home means on an ecological level. The farm, of course, is my "home," but that word has multiple layers of meaning and history. It means so many different things, depending on the context, the audience, the history of the word and the racial history of the person speaking.

⟶

In speaking and writing about land, indigenous people and the settler-colonizing people seem to have always stood on either side of a huge divide, an almost total misunderstanding even when they are speaking to each other using the same words.

All I have to do is read someone like Jeannette Armstrong, an Okanagan woman, the Wisdom Keeper for her Nation, and then compare her with someone who is saying similar things from a white perspective to see this gulf. Armstrong's words ring true because she is speaking from a place deeply rooted in her culture.

—∞—

As Jeannette Armstrong writes:

> The Okanagan word for "our place on the land"
> and "our language" is the same. We think of our
> language as the language of the land. The way
> we survived is to speak the language that the
> land offered us as its teachings. To know all the
> plants, animals, seasons, and geography is to
> construct language for them.
>
> We also refer to the land and our bodies with
> the same root syllable. The soil, the water, the
> air, and all the other life forms contributed parts
> to be our flesh. We are our land/place. Not to
> know and to celebrate this is to be without lan-
> guage and without land. It is to be displaced.[4]

Indigenous people have long been vocal about the
connection of their sense of self and well-being in relationship to where they live. As I both am and am not. I
feel at home in my place and yet my indigenous ancestors
were once Scottish. So I am white and a settler, yet I also
feel that my existence here and the existence of this place
are one and the same.

The split always there.

—∞—

But I do try because my home is essential to my experience and my existence. I am so much part of it that my

existence can't be separated from where I live. How would I live if I were pulled away from the farm and this valley? I would empty out like a sad balloon, like the frog that Annie Dillard wrote about in *Pilgrim at Tinker Creek*, sucked dry by a giant water bug so only the skin was left, floating on the pond looking alive but not. That would be me without the farm. And not any farm, not any place, this place. But why?

Because I committed myself to it at a young age, and because of that commitment, it taught me what I know?

Because it was, or became, both my mother and father? Because it fed me, sheltered me, plugged the gaping holes left in my psyche by school and life, with solitude and silence and a sense of belonging? It could just as easily be argued that I fixed my own psyche, that I projected this sense of belonging onto a place where I lived because I and it were simply here. But then why do I feel so strongly that I could not live anywhere else? I have lived other places, but not for long, and always knew I would come home again one day.

The land did more than teach me a sense of belonging as well as how to grow food. Everywhere I went, everything I did, became stories, layered over and over through the endless things done and not done, lives begun and ended, afternoons and evenings and mornings and weather, nights full of stars, nights full of darkness. Endless conversations, gossip, worry, laughter, neighbours, cups of tea and plates of cookies through seasons and years. Stories are the only way to make sense of it. Stories lay it down and then it becomes mythology.

Horses and cows, used to travelling in herds, find their home in the herd more than the landscape. Cats, who walk alone, are fascinated by any change in their landscape—even moving a chair is enough to engage their attention.

People find their homes and connect with their homes in many ways. But when humans lived as hunter-gatherers, knowledge of the land they lived in was essential. A people could not be separated from where they lived because they wouldn't know how to live, what to eat, where to find food, where to find stories, where to find identity. So to indigenous people, land, belonging, knowledge, identity, history and stories were all inextricably intertwined.

Today most North Americans, at least, seem to believe they can move easily but they experience this at deep and enormous psychic, emotional and social cost.

Australian writer Glenn Albrecht has taken this one step further in coining two new terms. He defines "solastalgia" as the kind of pain people feel when their environment is devastated. "Solastalgia is the pain or sickness caused by the loss or lack of solace and the sense of desolation connected to the present state of one's home and territory. It is the 'lived experience' of negative environmental change. It is the homesickness you have when you are still at home. It is that feeling you have when your sense of place is under attack."[5]

And he defines "soliphilia" as "the love of and responsibility for a place, bioregion, planet and the unity of interrelated interests within it. The soli is from solidarity with meanings connected to: A union of interests, purposes, or sympathies among members of a group; fellowship of responsibilities and interests."[6]

⸺

Writers have been writing about place, home, nature, for a long time. In North America, in the US, such writers as John Muir, Henry Thoreau, Aldo Leopold and Rachel Carson created a tradition known under the general and vague rubric of nature writing, but often dealing with a wide variety of subjects from a wide variety of viewpoints.

Modern "nature" writing is a rich and varied field that continues to grow and continues to be impossible to define. It also continues to be an area of contradiction, and richly diverse attitudes toward its subject matter. Attitudes and subject matter differ but all the participants are writers who respond personally and experientially to their relationship with what is around them, both non-human and human.

I don't consider myself a "nature" writer. I do write about my relationship with a particular place and its effects on me and to my place in this ecosystem. I am not enough of a naturalist to have an expert and technical understanding of who or what lives around me. I pay attention to who lives here, to coyotes, to ravens and eagles quarrelling overhead, to the swallows on my deck, the ospreys circling overhead when the dogs and I walk to the beach, and I try to see and understand what I can know of their being.

Most of all, I pay attention to what the land teaches me as I garden, walk, talk, think. Despite my years of education, all that I have truly learned comes from walking and thinking about where I live. About service to this place. About the history of farming and human-land interbeing. The knowledge that the land gives is through what I experience: sensory knowledge, working knowledge, physical knowledge, the memory of what has endured and what has changed. The understanding the land gives is through the lens it provides, a lens made from time and stories.

⸺

I am hoping to transfer my sense of love and connection to this place to my grandchildren. My parents grew or made or manufactured almost everything we ate. Even in the evenings, my mother knitted sweaters, or darned socks. My father made furniture in the basement or mended tools. He made all the handles for hoes, rakes and shovels, considering the bought handles useless things that cracked or broke too easily. He made furniture and fixed machines. My parents did these things with no consideration that they might be considered "creative" or "authentic" or traditional. They did them because these things needed to be done.

But I don't know how to transmit the reality of farming and a life lived outside to a generation that has never farmed and rarely goes outside. My parents and I stood on either side of a divide, but we had the land and work and a language in common. When I came home from university or teaching and sat at the dinner table, we talked the same language: weather, the garden, work.

And now, to my grandchildren, I am afraid I will become quaint, a relic, someone who gardens and farms as a strange and obsessive hobby. What I have for them are stories that carry the idea of family and land, of what men and women can do and how they can live. And more than stories, I have the land itself and the ability to invite them to walk and live on the land and to have their own experiences to layer onto the physical landscape.

Jeannette Armstrong says:

> I see that my being is present in this generation
> and in our future ones, just as the generations
> of the past speak to me through stories. I know
> that community is made up of extended fami-
> lies moving together over the landscape of time,
> through generations converging and dividing
> like a cell while remaining essentially the same
> as community. I see that in sustainable societies,
> extended family and community are inseparable.[7]

I know this actual embodiment of story is important,
both in terms of place (the farm) and in terms of how I
(like my father) personally embody the place. I also have
the presence of living on the land, which, whether I "teach"
them to farm or not, they witness and learn from.

And still I worry that if they don't have the experience of living here, how will they understand such stories?
Unless Louis someday builds a log house by hand, will
the story of his great-grandfather, while colourful, remain
abstract to him? Human beings need concrete experience to connect to understanding. To learn about outside,
Louis needs to hear the stories and then go outside. Or
go be with animals. Or garden. Or farm. Otherwise it will
become and remain mythology, which is of some use, but
ungrounded and abstract.

Perhaps that is why so many people get stories of
farming so wrong these days and why that woman in the
workshop seemed afraid of being trapped in a relationship she didn't want to understand.

Turning land into language and stories requires both
experience of a place and knowledge. It requires the names

that are held in that knowledge and experience, the names that adhere to it over time. It requires the sensory knowledge of a place over time, the infinitely varied and multiplying scents, colours, sounds, textures: the taste of fruit in summer and the smell of earth in winter and spring.

This is the one sure gift I can give as my grandchildren grow. When they ask, I can tell them stories of legendary times when their great-grandparents lived here and built a farm. They will look at me and the stories will nest in their memories and I hope, continue—the voice of their grandmother, telling them where they are and where they can belong.

Cake

I went to the library two weeks ago and got a bunch of cookbooks and made a chocolate cake. Everyone said it was a perfect cake. It disappeared in a heartbeat. It is true. I bake well. So now what? What shall I cook next to follow up this triumph?

And on the same note, we shall have yet another party at the farm in June for all the Gemini/Cancer folk all having birthdays clustered around this time. It will be a big potluck outside unless the mosquitoes arrive, in which case, we will huddle inside as if a nasty blizzard has arrived, with brief forays outside huddled in nets to water, weed, pick raspberries, feed things. Every year, I manage pretty well except for my ankles which swell and burn and are covered in bites and which I scratch all night so in the morning, my sheets are covered with red splotches.

The height of summer is the time we wait for all year and then it goes in a flash but it also seems to last forever. The garden stuffed with beans, peas, kohlrabi, little zucchinis, tomatoes starting to ripen. Soon there will be more and more, an avalanche of good food. There is something deeply ancient and satisfying in watching it pile up and then starting to put it away—not that anyone wants to give one tiny thought to winter, not yet anyway, but it lurks there anyway, the old cold beast.

And swimming—such pure pleasure—the beach, a book, a beer, a chair, an umbrella. I sit until I am roasting hot, then into the cold water, splash, roll, back to sitting. Summer is *sensuality and beauty, evening lightning on the blue shadowed mountains, sudden showers, long deep sleeps under a mosquito net. Flowers burst and bloom from every cranny, lavatera and white marigolds edge the garden, red zinnias beside the green-house. The morning bird chorus begins at 4 a.m.*

The Place Looks Back

NEXT DOOR TO OUR FARM on Kootenay Lake, yet another couple is building a summer home. These days, as the Alberta boomers retire, the price of lake frontage continues to climb. I don't know this couple, but I wonder what they think about when they look out at the picture their windows make. The land upon which they have placed this house was once a mossy tree-covered glade above the lake. I used to go there just to see the colours of the moss. Now it has been bulldozed, gravelled, levelled, staked, concreted—and underneath, unknown and unseen, the earth is walking, moving, breathing, being born, living and dying.

Of course, I don't know these people and I don't know what they see. In fact I don't know anything about them. In the summer, we will live side by side, ignoring each other like people in a too-crowded apartment building, even with a half-mile of space between us.

My father, who was from a different era, a different system and different values, would have gone over and introduced himself. He probably would have annoyed the man by telling him all kinds of stories about farming and the neighbourhood, and he would have ended by charming him. People were charmed by my father because they had never met anyone like him before and probably never would again. My father assumed everyone was his neighbour. When he came to see me in the city, he talked to people in stores or on the bus.

For him, neighbours mattered. Who they were and what they thought about things mattered less than the fact that they were present and available to be neighboured with.

But I'm a writer who imagines things so instead of introducing myself, I imagine these people looking at the pictures their windows make. Perhaps they say to their friends, "Look at the view," and the friends say, "Isn't it pretty," and "Aren't you lucky?" I know because this is what my friends say to me, and I often think how some of what I imagine might be true because before they built this enormous summer home, projecting out over the water, they put up a sign naming this place Wood Nymph Trail.

How odd that a view should be worth so much money. Lakeshore in our area is all granite, difficult to build on. Sewage has to be pumped away from the lake. But still people come, crowding in to build anywhere they can see, or even glimpse, the water.

People are buying a view, rather like the way speculators used to buy gold and diamonds. It's strange that a view should be worth so much money because it is useless other than as a picture. The world is beautiful, but that term is only human. Unlike an ecosystem or water or trees or other components of the world, a view means nothing unless it is somehow owned and appreciated by humans. Selling views is like putting the world in a zoo, carving it into pieces, and putting frames around the pieces—the world as art, framed on a wall.

When I was sitting with my daughter this summer on my deck, we stared out at the lake and the clouds rolling by. The deck looks out over a pond, green fields, the lake and the statuesque blue mountains, stoically eternal.

"It's like a very slow movie," she said. "It's so entertaining."

My daughter, who is a landscape architect and a much better gardener than I will ever be, then told me about the theory that humans want and need to be up high so they can see what and who is coming.

This theory, called "prospect refuge theory," claims that humans get "aesthetic satisfaction" from the "contemplation of landscape." According to writer Jay Appleton in *The Experience of Landscape*, this stems from the "spontaneous perception of landscape features which, in their shapes, colours, spatial arrangements and other visible attributes, act as sign-stimuli indicative of environmental conditions favourable to survival, whether they really are favourable or not."[1]

In other words, we like the view because it used to help us survive. Appleton adds, "Where he has an unimpeded opportunity to see we can call it a prospect. Where he has an opportunity to hide, a refuge... To this... aesthetic hypothesis we can apply the name prospect-refuge theory."[2]

Although this doesn't explain why we want so desperately to stare at water.

There's also a theory that humans were once cliff dwellers, another that we were once shore-dwelling sun baskers. Perhaps if we can combine cliff dwelling with water viewing, so that we can be sun basking cliff dwellers, we get a situation that satisfies some deep biological urge. And then if we can capture this behind glass so that we're safe, and the prospect before us is only a prospect, we can relax. The view is about aesthetics, not survival. And since the urge of modern day capitalism is to commodify everything, once something is a view, it can be sold off to an "owner."

But of course if you're behind glass it's difficult to know the other beings who are in this place with you. It also excludes the possibility of them knowing you.

⸺⸺

For many years, I have had people come to the farm, look around in wonder and pronounce it beautiful. After that, they often tell me how lucky I am. And although I smile and nod and agree with them, I am never quite sure that we really understand each other.

Because after that we usually go for a walk, and I find out they are afraid, variously, of mosquitoes, wasps, bears, cougars, spiders, lightning… They look at the garden, the fruit trees, the fields and the animals with interest but little comprehension, and then we go inside for tea. They mean well, but it seems that these wonderful, caring friends sit on another side of a cultural divide that, for me, keeps getting wider.

Lately I have been getting a plethora of thirty-something visitors, usually through an international program called WWOOF (World Wide Opportunities on Organic Farms). The WOOFers, as they call themselves, are immensely helpful and interested in the farm and also often in the latest fad about gardening or organic food or herbs, but they are often ignorant about where they have landed.

One of these young men asked me the other day if there were whales in the lake. "Nope," I said, with some amused despair, "never seen any."

What are they seeing, all these varied people, and what do they think they are doing here? I have decided, after long experience and some thought, that perhaps what they are seeing must closely resemble a photograph and what they are doing is living, temporarily, within that photo. Or perhaps sometimes it's a painting. It doesn't matter. This still doesn't explain the vagueness of the term, "beautiful," or why they seem to think they are complimenting me, as well as the place, by saying it. To some extent, it is because of the quality John Berger in *Ways of Seeing*, terms, "glamour."[3] Glamour, he says, can't exist without envy. Since I, in some mysterious way, now "own" this expensive view; then somehow it is a compliment to me and gives me glamour that I live in such a beautiful place, that I am so "lucky" (which of course I am), and of course, immensely privileged to live here. So yes, of course, I also understand the edge of envy in their voices.

But there our communication tends to stop because although I hear what they are saying, I don't think this way about the farm, nor are our experiences parallel. After all, most of what they know about nature comes from pictures, movies or art, and most of what I know comes from experience. So there is a gap between their and my understanding of what I am doing here.

Sometimes I feel like I stand on the other side of this gap, a vast canyon, yelling gibberish to people on the other side. I want to explain that experience at the farm is interactive, that engaging with plants and animals takes time and practice.

I don't want to be too hard on my visitors, who are my friends and often even family. And really, how can my well-meaning and caring visitors know where they are or see what I see? After all, their experience of nature is that it is something "outside," outside of the house, outside of the city.

The non-human world, through sheer lack of opportunity for contact, is, to most people these days, unfamiliar. In the media, it is portrayed as something one looks at, briefly, or takes a picture of, or uses as a pretty toy—like the summer people who come for two weeks every year and drive around and around in circles in their boats or Sea-Doos.

All of these people, if asked, would say they love nature. They love animals, they value peace and quiet, they love the fresh food from the garden, they love the beauty of the farm. They are glad to get out of the city for a while, however briefly, and they are happy to have the experience of being here with me. But it's all about how it makes them feel. I don't denigrate their experience and I believe them when they say they love it here. What I don't understand is what they love.

No, that's not true.

When I was a child, living where I still live, I was endlessly swept up, transported and thrilled by the beauty of this place. I tried to think how to respond because such beauty seemed to call for a response. I began painting and drawing. I wanted to capture what I was seeing with paint and canvas, but eventually I realized that was impossible for me and I stopped painting.

So then I just walked around, looked at it, listened to it, smelled, watched it. I still do. But when I look out from the deck of my house these days, what I often see—despite my almost sixty years here—is how little I know or understand. Everywhere I walk there is an endless and intricate exchange of information among the billions of non-humans that exist here with me. An insect, a clump of grass—what do I really know about its interbeing and how it functions?

I know this is not a picture and that "being" here is an experience of being within an endless number of profound and complex relationships. I feel the depth of my not-knowing, my lack of ability to get outside of my humanness and begin to understand who or what is looking back at me. But because I have to find my own way to this relationship, I am the confused one, the lost one, the one still learning how to behave.

Or, as Gary Snyder says in *The Practice of the Wild*, "Our relation to the natural world takes place in a place, and it must be grounded in information and experience."[4] Although I have both information and experience and am always on the hunt for more, I feel I am just beginning. True, I was seduced into the relationship by beauty. But that's the beginning of understanding, not an end. I also worked on the farm from a very young age, and understood from listening to my parents that my work was necessary to our very survival as family. Or as my father put it, when I complained, "You work or you starve." Doing the work, I learned to love it.

Perhaps that can be true for people who come here too. I hope so.

⸺

My sister is a horse trainer and whenever she is around, we talk about horses, animals we have loved and ridden and danced and smooched with all our lives. And they're still a mystery and a wonder to both of us.

I like to talk to people about horses because many people who are unfamiliar with them are afraid of them. I was when I was a child because they were big and powerful, and even though I rode all the time, no one ever explained how to do it well and how to communicate with my horse. But my sister, who has also been riding for fifty years, has been learning from her horses all along. She loves her horses but, as she says, she's not nice to them. She's the lead mare and in charge and when she's around, even I can see that the horses are relieved and glad that she knows what she's doing, and that they are part of her herd. But they teach her as well, all the time.

I'm not particularly nice to my dogs either. They're working dogs and have jobs to do and that makes both them and me happy. I get to be the pack leader because I have the house and the food. It's a useful and ancient agreement, and they and I both honour it.

What they and I like best is walking around. My dogs are collies. They believe in walks. In the afternoons, if I haven't shown any signs of moving, they start, very determinedly, figuring out how to herd me out the door by staring at me, standing at the door, watching my every move until I get it.

I am never quite sure what it is about walks that is so important to dogs, but walking satisfies something in both of us. The dogs know their territory intimately by smell and sound and touch and taste. I know it somewhat more dimly, by looking around and listening. As Gary Snyder says, "Walking is the great adventure, the first meditation, a practice of heartiness and soul primary to humankind. Walking is the exact balance of spirit and humility."[5]

After our walk, after chores are done and the farm has settled to sleep, then they settle in on the rug for a snooze; we snooze together, in our dog-smelling warm den.

I would like to have similar clarity in my relationships with the other less-seen inhabitants of this place. When I go walking, it is often more like wandering. I stop, I start, I stare, I listen. This kind of walking is one way of greeting the non-human beings where they and I live.

"The world is watching; one cannot walk through a meadow or a forest without a ripple of report spreading out from one's passage. The thrush darts back, the jay squalls, a beetle scurries under the grasses and the signal is passed along. Every creature knows when a hawk is cruising or a human strolling."[6]

One of the moments I loved most as a child was the passage from the open fields into the trees, and the moments when I would stand inside the tree line, listening to my passage being announced, squirrel by squirrel, and raven by raven. I was very young when I figured out that the forest was watching and listening.

The human supposition that we are in some way superior to animals, can use them, impose our metaphors, ideas, experiments, emotions on them, is deeply entrenched in our cultural, not scientific, assumptions.

Our notion of human superiority is a supposition based, in part, on the fact that since we can't speak to animals, they must be stupid. The same supposition was used by colonizers all over the world. It's similar to the supposition that religious people make when they assume, without ever being able to prove it, that they not only know what God looks like, they know what and how he thinks.

I have a suspicion that animals understand us much more than we understand them. I laugh at myself every time the bear comes to eat "my" apples. But I still get mad—which is silly of me. The bear needs those apples, needs the fruit on the trees, needs what he or she sees as necessary to his/her survival. How can a bear tell that the wild apple tree down the road, also covered with apples, is any different from the one in my yard? On my part, I want to have some way to tell the bear not to smash the tree down, which to him or her is the easiest way to get at the apples, and to leave some for me. We're still working it out.

⸺⸺

This summer I had a lot of young Quebecois visitors. At first they thought the swallow nests above the deck were kind of cute, until they realized there were several wasp nests up there as well. I pontificated that wasps have a keen sense of smell and neighbourliness and would get to know them. They didn't believe me, but since it was my house, they couldn't do much about it. And after a while, they got used to sitting on the deck under a bunch of wasp nests, but it made them uneasy, not only because of the wasps but also because of the manifest craziness of their hostess, who made them sit there.

In fact, the chair on the deck in which I sat every morning to drink my coffee also had a wasp nest underneath it. I didn't really notice until my son sat in the same chair. He was indignant and hosed the chair down until all the wasps went away. Then the wasps were probably somewhat indignant, or at least I was on their behalf. But they appeared to accept it as some kind of natural disaster. When my son was gone, they rebuilt the nest in the chair and we went on having coffee together. What makes this a somewhat saner anecdote is that these were paper wasps, not yellowjackets. Paper wasps are a European import, and far less aggressive than yellowjackets.

Later in the summer, I went on a brief retreat by myself to a nature camp on a peninsula in the lake. I broke the rules and brought in a can of salmon to this vegetarian camp.

I sat on a stump and ate my sandwich while the wasps swarmed me in numbers that got a little overwhelming. I laid my plate down for them covered in bits of salmon and juice, and they literally licked it clean. But even while I was eating and being swarmed by wasps, I felt no sense of fear. They watched me devour this giant sandwich and never interfered except to buzz by my head in a constant breathing desire for food.

I walk through this world as both prey and predator. Mostly predator. Chris Irwin, the great horse trainer, says that horses know immediately when you enter their field that you are predator. If you turn your body and eyes towards them, their immediate instinct is to run. Since most people come into a field to catch horses, the horses' instinct is obviously correct. But if you politely turn your hips and your eyes away from them, they can relax. Such small communication allows a bond of understanding to begin, but only to begin. I practised such politeness walking by the wild ducks that settled in our pond this spring. Turned my eyes down, didn't stare, didn't stop or point. No predator behaviour. The ducks watched me, and then relaxed and sailed around the pond.

The place looks back. The place has thousands of eyes. The place has morality, courage, a culture, all within a pattern that lives and breathes and dies and is born, and within which I make my own interruption, my own breathing space. I watch the plants that grow along the road—another interruption, a clear space within a crowded pattern. Each finds its space in the light so that although there is crowding, there is lushness. The thimbleberries produce and produce, the blackcaps twine their way around and through and underneath; meanwhile the baby maples are putting their feet down, getting ready to make their own way.

When people get out of their cars in my yard in the summer, they look out across a grassy windblown pasture to the blue bulk of Steeple Mountain and a line of Selkirk Mountain peaks marching north. There are flowers and birds. In the summer mornings, unless it is pouring rain or too thick with mosquitoes, I take my coffee outside to the deck. I stare out over the glowing green fields. There are wasps and usually a few mosquitoes and hundreds of swallows. At that moment, I am always conscious—I remind myself to be conscious—that I am here, that it is beautiful, that I am unbelievably privileged and fortunate, and that I love being here. In *Ways of Seeing*, John Berger says two things that I try to be aware of: one is that how we see things is "affected by what we know or believe." The other is that the relationship between what we know and what we see is never settled.[7]

I wonder how my neighbours in the glassed-in house to the north of me are affected by what they see through their windows. I wonder if they come out of their door and see the ospreys whistling above them and say hello. I hope so.

For my own part, I may still be walking through this landscape I know so well somewhat blind and deaf but

I am always learning. I pay attention, I say hello. Each day I try to walk myself a little deeper into the reality of this living, breathing landscape, of which I am a small and sometimes lonely participant, somewhat aware, at last, of seeing and of being seen.

If All Months Were August

If all months were August and each day were sunny (with just a little rain at evening) and every moment were another slow, sun-exhausted, cool-water moment, I could lie under a beach umbrella, the dim blue shadow of Castle Mountain in the cobalt blue water, a distant osprey, a turquoise dragonfly rattling on my knee, robins flying over the bay, and small minnows flashing fins in the clear moon-shaped bay.

Oh, if only every month were August and school were always far away and books were nonsense and only wasps and ravens and bears made sense. If all months were August, I would live on peaches and corn, green apples and ripe tomatoes forever. I would eat peaches for every meal and never get tired of them. If all months were August, work would be a distant worry. The sun would slant over yellow fields and break my heart with nostalgia but it wouldn't matter because it would still be August and winter only a memory. Summer is for always: the hawk on the cedar tree, blue herons in the pond, and me sitting on the deck at night under a full yellow moon, with only an occasional meteor to mark the passing of time. If every month were August, I would never get cold or tired. I would swim every day in my warm brown skin and never shiver. I would walk home in the hot sun, and the moon would rise, full and ripe over the Purcells, and the night would be *full of crickets. I would sit on the beach by the fire, staring into the many-coloured flames, then walk home under the full moon, the dogs running ahead, to warm beds and warm quilts, and I would lie down to an untroubled sleep and a bird-song waking.*

The Sudden Falling of the Light

THERE IS A MOMENT EVERY August when I wake up and the light has changed. When I walk out to drink my coffee on the deck in the early morning, the hills across the dark lake are shaded in a genteel navy blue, and I know the light is falling, down toward fall. Gone overnight is the sturdy growth light of early summer, when all was optimism and first blooms, when dandelions, shasta daisies and buttercups speckled the hayfields, and peas, beans and corn pushed out of the dirt with fierce energy.

When the sun rises on a late August morning, light slants over the great blue-green fan of mountain behind the farm, so that every tree stands out. The light is clear. Everything springs forth in blended colours of green and pale blue with brown undertones. The light slants through the morning clouds on the mountains and breaks my heart. Summer is over, summer is over, it cries. But the sky is still blue. It's still hot. Every afternoon, I walk to the beach through the gold- and purple-tinged fields, where the wind bends and ripples the timothy hay, heavy with seeds. My feet plod through dust, small spirals, the dogs run and run, happy to be going to the beach, to be going anywhere. And then we sit together through the long golden afternoon while the light falls and falls, and when we go home the garden is heavy with food, with corn and tomatoes and potatoes and squash, big lumpy chunks of food, fall food, full of nourishment and protein, demanding to have something done about it.

Every year, this changing of the light surprises me. I know it's coming, but summer is hot and bright and eternal. And then the light bends toward fall. Everyone, everything quickens. The conversation at store counters and over the phone with neighbours is of school and firewood and making apple juice. Talk is of harvest and storage and canning. The squirrel yodels a challenge from the walnut trees. Bears prowl the mountain, plotting their nightly forays and forages into the ripening fruit on the orchard trees.

I resist. I avoid. I sit on the beach because summer is always forever; then I go inside in the sudden dark evening, and all the golden days are a lie, because the light has changed.

The summer is forever and so is winter. Both are liars. The season swings around light like a child swinging on a pole, blonde hair flying.

—————

My dog Kin is old. He's a black border collie-lab cross, who came to me as a puppy. The first night I had him, he flopped down beside my bed with a sigh and went to sleep, already committed to being my dog, and thus, at home. Now he's fourteen and deaf and crippled. What hurts is that he won't surrender to being old, nor do I want him to. We both have arthritis but his is worse than mine. When the other dogs run to the beach he runs too, slow and stiff, but running. He comes home from these walks to the beach, flops down and sleeps like he's dead. After I feed him in the basement, he stands at the bottom of the stairs panting and whining until I go down and let him out the basement door and take him the long way around so he can come in the front door.

When I see him now, I see myself. I straighten my shoulders and try not to limp. When he sleeps, sometimes I bend and touch his head and say don't leave me, don't leave me. He has been my shadow for fourteen years. His death is my own, but faster.

The last time I left the farm and lived in a city, I was over fifty, back at school as well as teaching—too busy, trying to accomplish too much at once. One sunny spring morning, when I left my house to walk the five blocks to the bus, very suddenly, a habit fell off my shoulders like a crust of dried mud. It was a weight I hadn't known I was carrying. Why did I care, I thought, why did I care what other people thought? What anyone thought. All those years of worry and ambition and striving and measuring up and worrying about success, about failure. I was walking by the Ukrainian cathedral with its stained glass windows of the Virgin and her child. The leaves on the trees at the edge of the parking lot were brand new, bright and shiny.

There was a whole flurry of ravens on the network of phone and electric wires above me, screaming raven conversation. I stood and watched them. The ravens were having a long and complicated conversation, and I was falling down and up at the same time, falling toward my elder age, toward not caring. While I stood there, bits of other habits fell away, worry about looks, behaviour, success, getting ahead. Shards of ambition flaked onto the ground like bits of silver off a mirror.

Now, I thought, I could wander all day talking to ravens and no one would ever care. But I went on to the bus stop, so much lighter than I had been before.

⸺⸺

When I was a child on this farm, one of my jobs was picking seemingly endless rows of raspberries and strawberries. Picking berries was only about trying to get finished in order to get to the beach. Whenever we thought we were done, my father would come to inspect and of course he always found berries we had missed. If someone had tried to explain to me then that this experience was good for me, they might have ended up covered in squashed fruit.

This year, I was picking late summer blackberries with my friend K.L. She had just come from a retreat centre where she and I had organized a workshop on how people connect to where they live. But this year I hadn't gone because I was teaching and working.

"I asked people at the Sense of Place workshop to talk about their peak environmental experience," she said. Hers, she said, was picking berries, and in the hot late sun, looping the vicious strands of blackberry vine out of the way, sidestepping and trying to squash down the four-foot thistles to get into the middle of the blackberry jungle, filling bucket after bucket with berries, I understood what she meant. Time tends to slide away while picking berries; it's all about concentration, pursuit and repetition.

"I don't know if I've ever had a peak environmental experience," I said finally. "Living here, it's all pretty peak. Maybe it would be dying on the beach at sunset…"

"I'm Estonian," she said. "Berry picking is what we used to do to survive. This is what I love." I thought about the First Nations people I used to work with, telling me about berry picking and why it mattered to them. A blackberry vine tried to strangle me; bits of thistle were lodged under my toes. Our conversation continued but it was garbled as we ducked and wove our way deeper, trying to get those last blackest, ripest berries, always just out of reach. And finally, eventually, we crawled out, laughing at ourselves, two happy-crazy women, covered with scratches and mosquito bites.

⁂

I stand in the garden in the middle of October after the light has toppled over into fall. The poplars and larch are bright yellow, the blue mountain is shadowed and soft, and I grieve this year's garden. The tomato plants are shrivelled, the zucchini leaves have collapsed. All that's left to harvest are a few carrots. It's silly. I am grieving nothing, it's over and gone, it's done and next spring it will start again. Every garden is the same, every garden is different. Every year, some things grow better than others. It's an adventure, a journey, a mystery and a complexity.

The garden in fall is a muddled jungle. Everything has been rushing to completion so the weeds that are inevitably left over are huge, putting out seed heads fast, and will do so until they freeze. My once neat garden is now a tangle of weeds, cornhusks, squash vines, giant rotting zucchini that even the pigs won't eat. My job is to gather it all up and chuck it on the compost. My job is to eat the last few tiny golden tomatoes still hanging on the dying plants. My job is to laugh at myself for being so foolish and soft and sentimental, for loving a garden and grieving its end.

⸺

It makes no sense to grow old, to lose my eyesight, to have stiff knees that won't bend, to walk through the world with my body slowly bending over, like the ancient pine tree that has been growing out of a crack in the granite over the lake for as long as I can remember. Every year this tree has bent a little farther toward the water, defying gravity and the odds. It has one long root stuck in the granite, holding on. It's only possible, as I age, to weigh and assess the good and bad, the light and dark. It's only possible to walk through the world watching it shift and change under my feet, before my eyes, watching the eternal and the present mix and churn together. When I visit with my friends of thirty years, I wonder occasionally, which one of us will see the other leave.

⸺

When I went to bed one September evening, the radio was playing the saddest music of all, September songs. It was three weeks after the light had changed, and finally I was alone at the farm. All the hustle of the summer was done, the visits, cars in and out of the yard, picking the abundance of the garden and turning it into enormous dinners, good food and wine and friends and family.

This summer my youngest son got married and we prepared dinner for eighty-seven people. Music, dancing and laughter continued until dawn. When we were done, everyone said that was so much fun, let's do it again next year without the wedding. Maybe we will. It was so good to celebrate summer, to celebrate the garden, harvest, and family, the four, no, five generations that have called this place home.

After the party, I lay in the dark, slightly anxious as I always am before sleeping. With no defences between myself and the music on the radio, the years and the experiences and the memories I have lived through came in a flood. Perhaps it was something similar to what people report from a near-death experience, but I wasn't dying, merely lying at the bottom of the enormous tottering stack of things I have done, the people I have loved and missed, the children and dogs and cats and weather and time, and the winter coming in, coming in again.

It was like being at the bottom of a well, looking up at the stars, or being inside the image I sometimes have of us all walking around in this enormous soup of air and atmosphere, balancing columns made up of tons of air on our heads—only I was balancing my whole life. Fortunately for me, I was tired and the radio stopped playing sad music and I drifted away to sleep, carefully holding my life, like a stack of fragile, blown-glass balls, safe in my arms.

Summer's End

The farm feels contented. The twenty pigs have mud and grass to roam and roll in. The gardens don't need irrigation because there has been a rain shower every afternoon. Plus there are suddenly people everywhere, my sister, my son, my son's girlfriend, my grandson and his friend. Tonight, I made dinner with every available garden vegetable—cucumber soup, a big salad and a big veggie stir fry and then raspberry cake for dessert.

What a summer ride it was. And not quite over yet. But we are sliding down the far side of it toward September, school, classes, harvest, chilly nights, the end of bugs.

It has been a summer of many comings and goings and arrivals, and many melodramas, none of which particularly concerned me; they just happened in my house. But right now I am about to be invaded by an enormous horde of tomatoes, and tomorrow morning I will make a winter's supply of vegetable soup stock to tuck away. There are also eggplants, corn, peppers, cabbage to process. The fifty enormous broccoli plants that never made a broccoli head can go to the pigs. The happy, happy pigs, galloping about their pen, lolling in their creek, eating and eating… visiting with their many admirers.

My brain is waking up from its summer's rest. I am finally, reluctantly, emailing my students to let them know the dreaded day approaches when we all have to get our arses back in chairs in front of our little glowing screens and forget all this touchy-feely running around in nature/gardening stuff.

I had a day almost alone today in which to recover my life and write about it. One neighbour came for tea and my brother came to drop off some stuff to store in my already crammed basement. But other than that, it was a fine and quiet day, something I need more than breath. It was almost cold, almost raining, lovely grey quiet mist hanging in the gaps between the mountains, gold light at evening, night coming down so fast now, almost dark by 6 p.m., the dogs sleeping on the rug and the cats curled up on the bed.

Last week was full of people. Just when I thought I was done and everyone had left, the house filled up again.

I sit for only a moment. Delicious sitting. Delicious not-thinking. The trees have turned yellow on the mountains across the lake. I want to soak in silence forever. Finally, I hoist myself to my feet and pick a wheelbarrow of giant zucchini for the pigs. The fruit finally, finally, all done. Last week we made apple juice, my friend Yvette canned peaches, pears and prunes for me, I have a freezer full of stuff, I just need some more wood for the stove and I will be ready for winter.

Writing After Nature

THE YOUNG WRITER LEANED IN closer. "Do you really kill your own animals?" he asked.

I explained that yes, we do. He stared at me. We were sitting at dinner in a Vancouver restaurant, with a group of other writers, just prior to a book launch of a non-fiction anthology I had helped to edit.

"I'm an urban man," he said with a shudder. "I'm writing about how cities can make people happy." I thought that was interesting and said so. I talked about my neighbourhood in Vancouver and how much I had enjoyed living there, how much more convenient it was than the farm. And then it was time to go.

We walked down the street together toward the Vancouver library while I thought of all the other stories I could tell this man and would never get a chance to. I wondered how long it would take us to really understand one another?

⸺◈◈◈⸺

Because our farm was on the edge of wilderness, because my father was a subsistence farmer always fighting with weeds, weather, coyotes, bears, ravens, hawks and owls;

and because, in contrast, my mother loved nature and loved flowers, gardens and animals in particular, I grew up with a contradictory and conflicted view of human-nature relationships. My brothers and my sister and I spent our time when we were not in school either working on the farm or exploring the woods and lakeshore around the farm. And because, as a child, I lived far more intensely in my relationships with animals than I did with people, I was never quite sure where my loyalties should, or did, lie.

～

My father assumed we kids would help with killing animals for meat. I would stand right beside him as he shot the pig or cow in the head and then cut its throat so there wouldn't be blood left in the meat. This didn't bother me because in an instant the animal changed from being our pet to being dead, and a dead animal looks so different from a live one.

I may have been accustomed to killing our farm animals, but I always found the idea of killing wild animals disconcerting. Our neighbour, Wally Johnson, the trapper, made his livelihood from his knowledge of animals. In order to track, trap and kill animals, he had gathered a broad and expert knowledge of place, wilderness, animal habitats, animal behaviour and survival. Throughout the winter, he would snowshoe from trapper's cabin to trapper's cabin, living on the meat of the animals he trapped or shot. Wally was a man of his time and understanding. He loved birds in particular, knew as much as he could about them. Everything else, especially predator animals, he believed should be exterminated.

Even at the age of seven or eight, I argued with him about it. And Wally, to his credit, listened carefully and gravely to my arguments and then argued back. I was so fired up by my debates with Wally that in grade four, I wrote an essay titled "The Balance of Nature" to encapsulate my arguments.

But these days, I see examples of change, resilience and balance in nature around me all the time. For example, new residents and visitors often perceive the Kootenay–Columbia region of the world as a somewhat peaceful near-wilderness. But in fact much of the fish population in Kootenay Lake is artificially maintained and one of the former main fish populations—salmon—is gone. The main rivers all have multiple dams on them, the forests have all been logged at least once and sometimes twice, and many of the animal populations are at risk. Now, environmentalists fight to keep giant ski resorts off the glaciers, methane drilling out of the last few undisturbed areas, and private power projects off the smaller creeks. Because I have now lived here for sixty years, I have watched how our relationships with the place shift as the nature of the community shifts.

These days, I write stories and poems and I go for walks and write about the ideas that emerge while I am walking around. I also read books about place or nature or animals. And if I start from walking around, from looking at what is going on with the people around me, as well as animals, both domestic and wild, birds, insects, plants, trees and the interrelationships among them all, then the questions that arise connect me both to the local and the global—how we live, ethically, here and elsewhere. Thus, I am always both in a place of discovery and a place of familiarity.

The contradictions in how both I and other people interact with this world continue to puzzle, fascinate and confound me.

Animals never lie, for one thing. They always have a good reason for their behaviour—though it's often difficult to see because they are reacting to situations much differently than humans do. And because animals are usually reacting to human behaviour, they often act as a mirror for people who can choose to look or not. A bear swinging on the end of a plum tree branch in order to break it off and bring it to the ground has no idea that he is smashing a tree that I believe I have some claim to. A horse reacting in fear to being hurt by its rider is trying to escape pain and the human it sees as a predator.

Our lack of understanding of non-human life is constantly being made sharper and more poignant both by the growing environmental crisis in the world, the resulting growth of environmental awareness, particularly in the sciences and some other sectors of humanity, and the changing nature of human relationships with the non-human world. In fact, as warnings of environmental problems continue from many quarters, the stakes in such relationships get higher and higher. These environmental problems, if you are paying attention, will scare you into the heebie-jeebies: global warming, water scarcity, energy shortages, polluted oceans with giant floating islands of plastic, and on and on... and even if many of these effects can be mitigated by environmental efforts, there is still enough there to give any conscious person, especially those of us with grandchildren, long pause for thought.

And some consideration of the irony of it all. After all, what will be the vaunted philosophy of humanity, the

loftiness of the human story and the illusion of our superiority, if life on earth as we know it is undone by something
so powerful and mundane as the weather?

Or, as Kate Rigby, an Australian writer who has done
of lot of thoughtful writing on this subject puts it: "... the
tidal wave of extinction that such anthropogenic factors is
now engendering surely threatens the particular *oikos*, the
planetary community of living beings into which humanity was born."[1]

Such warnings continue almost non-stop these days
and in the meantime, very little changes, or seems to
change, in the modern way of life. So what then is the role,
or should, or could be the role of writers in general, and
non-fiction writers in particular, who want or choose to
write about the non-human world?

Rigby continues:

> We are going to need the very best science and
> the greatest technical ingenuity that we can
> muster both in moving towards a post-fossil-
> fuel economy and in preparing ourselves for the
> potentially catastrophic climate change impacts
> that are now already inevitable. However, climate
> change is not just a technical problem requiring
> a technical "fix." Both in its causes and effects it
> is also a socio-economic, political, cultural, and
> ethical problem.[2]

And I would add, are environmental issues in general.
And that is where the writers must, and I hope, will, start
to weigh in, in bigger and stronger numbers than they
have done so far. All through history, writers have illumi-
nated, examined and posited the themes, the tropes, the

attitudes, the interpretations and the assumptions within human/non-human interactions. And now it is a necessary and crucial task to revisit this area of thought and ask ourselves what is worth keeping, and what needs to be re-thought, and what needs to be written.

—◦◦◦—

Canada has a long and venerable history of nature writing, from Sir Charles G.D. Roberts, Ernest Thompson Seton, the iconic Susanna Moodie and Catherine Parr Trail, to current Canadian writers such as Trevor Herriot, Jan Zwicky, Don McKay, Tim Lilburn and many others. Canadian writers have taken on nature, in part, I think, because for so long there was so much of it and so little of us. I mention these four current writers in particular because to me they are writers who are writing from a shift in perspective and moving past many of the standard tropes of so called nature writing.

In non-fiction nature writing the most common narrative is the urban person who leaves urbanity and has a journey of discovery in some rural or wild area, often meeting animals and quaint rural characters along the way and then returning to his or her real life somewhat wiser and refreshed by his or her sojourn in "nature," which is often portrayed as beautiful, spiritual and educational; although, of course, there is also another type of story in which someone has a terrible encounter with some terrifying or traumatic aspect of nature: being attacked by a bear or getting caught in an avalanche or getting lost or having some other traumatic experience and returning changed from that as well.

Many threads weave through such "nature writing": reverence, fear, sentimentality, pomposity, spirituality, a sense of wonder and a quest for understanding. But nature writing has rarely been seen as political or economic, or engaged in shifting social and cultural paradigms. But that is exactly what is happening now. As we create this "new" nature writing, we shall have to decide what to call it. Because even the term, nature writing, has become problematic and defining the genre is becoming increasingly difficult.

⸺

Zoe Landale and I came up with this definition for a non-fiction anthology that we edited:

> Good nature writing incorporates clear and well-researched information about the natural world; at the same time, it also delineates deep personal involvement and philosophical premises. Nature writing calls upon its readers to re-examine and revalue more-than-human beings, places and histories. More and more, nature writing involves ecologically oriented interests, an examination of ecological identity, of self in nature. The last decade has seen the emergence of an ecocritical perspective where it is equally important to emphasize both language, and the complex unfolding of life on earth. The term nature writing is problematic. But whether we call it nature writing, environmental literature, ecopoetics or something else

entirely, the question that such work funda-
mentally addresses is that of the relationship
between humanity and other-than-human
beings, their places and histories.[3]

More and more, writers and environmentalists are
becoming aware that making a distinction between
"nature" and "humans" is a false dichotomy. As that venerable sage Gary Snyder wrote in 2007, "What we refer to as
nature or the environment or the world is our endangered
habitat and home and we are its problem species."[4]

So what then does this urgent vital and alert writing
consist of, and how do we recognize it?

⸺

Canada was settled by people who contended with nature
and who shared a pretty clear set of values that they got
from John Stuart Mill and others such as Adam Smith,
values that prioritized the idea that land was property
and that exploiting it in whatever way was possible and
necessary—digging up, cutting down, damming, min-
ing, etc.—was the right way to create wealth and security.
And because the world seemed unlimited, very few people,
then, ever questioned this idea. In addition, I don't think
anyone questioned the idea of the innate superiority of
humans over animals.

Many of these ideas are now being questioned, in particular, the idea of limitless growth where land and animals
only have value as economic resources, but there are still
all sorts of other mainstream media ideas that rarely get
examined or deconstructed: the idea of scenery as a commodity or artifact, or of nature as some kind of blank pal-
ette upon which we can project our fantasies, with animals

as clichés of themselves, usually alternating between savage and cute; and the endless, subtle idea of nature as something that can be owned, controlled, exploited and made safe. But now, of course, the idea of resource exploitation has smacked directly into what the Club of Rome (a global think tank founded in 1968) termed, in 1972, "the limits to growth" and the various warnings about problems created by the ways that humans live in the world.

So given these conflicts, does the term nature writing still have any meaning? Can we write about "wilderness," for example, when by definition it is beyond words and when, as American writer and environmentalist Bill McKibben has essentially said, there is no such thing because there is now no place on earth where the impact of humanity hasn't been felt?

In fact, one of the great difficulties of nature writing is that not just the term itself but so much of the language around it is problematic. The English word "nature" is from the Latin, *natura*, meaning "birth, constitution, character, course of things." *Natura* comes from the root word, *nasci*, to be born, or the root word, *nat*; so nation, natal and native all share this root term. In other words, nature means everything that exists: the rural, the urban and the wild. Wild is a somewhat more useful word. It essentially means uncultivated and undomesticated. Wilderness has a variety of meanings; from the OED: "a wild and uncultivated region, as of forest or desert, uninhabited or inhabited only by wild animals; a tract of wasteland, a tract of land officially designated as such, any desolate tract, as of open sea; a part of a garden set apart for plants growing with unchecked luxuriance; a bewildering mass or collection."

So these cultural assumptions are already built historically into the language. Wilderness is both desolate and bewildering. But only to those who aren't "wild," which of course, by definition, means humans who are both domesticated and cultivated.

⸺

And there is of course another parallel language problem, which is that because we have language, and as far as we know, or assume we know, the non-human world does not, will we also inevitably end up speaking for those who don't speak, or at least not in any language we yet understand?

But my question is, if we, the writers, don't begin to unpack these complexities and conflicts, who will? The environment today is too often seen as the province of the technicians. Biologists and ecologists and hydrologists have an enormous fund of knowledge of how things work, but this requires translation, collaboration, interpretation, analysis, and most importantly, it requires the addition of story to make it come alive and be relevant and understandable.

Non-fiction writers don't write specifically about ideas, theories or philosophy. What we do is embed both ideas and information in stories, research and language. As our understanding of how the non-human world grows, and its ecological, environmental and economic issues are knotted together, there is also a growing collection of philosophical and ethical issues that writers can't necessarily solve but can—and I think should—address, not as philosophers but as storytellers.

For example, there are laws and customs and cultural mores and shared understandings about relationships between people. There is general agreement on how people ought to treat each other, even when they don't; and when they treat each other unfairly or violently or viciously, there are laws to intervene. And now, more and more, there are shared cultural understandings about representation, about how we can talk or write about people seen as marginalized, dispossessed or non-mainstream.

Granted, conflicts between people are the central and necessary ingredient in our fiction and non-fiction, and in general, the resolution of these stories is such because reader and writer agree about the values inherent in the resolution; that is, we recognize justice and retribution when we see it. We recognize love and courage and transformation in humans.

But do we have, or can we have, ideas of what represents justice, love, courage and redemption in writing about the non-human? Can there be such a thing within a conflict between the human and the non-human, the "wild" world? And what about other aspects of it that people find so inspiring, as well as frightening and terrible? How do we write about those without falling into clichés of description, standard tropes of human and non-human behaviour, too familiar images? How can we get outside of our human assumptions about creatures without a language that we can interpret?

Interestingly, in many universities, philosophers, social scientists, literary critics, ethicists and many others are having a fairly intense discussion about all these issues. And for biologists, such ideas as language and intelligence, and how they, in particular, and we, in general, interpret the meaning of such terms, tends to be an interesting preoccupation. But until humans somehow manage to find a way to interpret and understand animal language itself, science is stuck with inferring and decoding behaviour and ecology within a human framework.

As are writers. In fact, it is impossible to write about the actual non-human world, but only about human representations and ideas of it. The more closely such representation is grounded in observation, knowledge and actuality, the more accurate it becomes. But it will always remain a portrait, an image, a map, not the thing or the place itself. Or, paraphrasing Dutch writer Cees Nooteboom, speaking on CBC's *Writers & Company*, March 21, 2010: whenever we write about feelings in nature, we are writing about our own feelings.[5]

And thus, stories and writing about the non-human world inevitably come to have a huge political—as well as a social, cultural and, ultimately, economic—dimension. The politics of representation are indeed complex, and yet we have faced similar issues. In the last forty years, for example, complex discussions about cultural appropriation, racism, class issues and feminism have disrupted and changed mainstream discourse. The last forty years have also been a time when sectors of humanity who were previously almost completely unrepresented in literature have spoken and written about their own reality, a process that is still contributing to our understanding of each other. The ability of other previously marginalized groups—women, First Nations, people of colour, immigrants, third world peoples, gay, lesbian, bisexual and transsexual people—to break this barrier of silence and speak for themselves has had a huge and transforming impact on the social and cultural understanding of humanity. But the non-human world, at least so far, does not have this autobiographical ability despite a constant and ongoing effort by many researchers to either teach animals to speak or to decipher the noises of animals that might appear to have language, such as dolphins or whales or parrots.

As far as we know, the non-human world is dependent upon humans to represent them fairly and well. Whether this can be done or not, continues to be an amazing and deeply engrossing (at least to me) challenge.

There is another aspect to this issue. Australian writer Kate Rigby describes it as "repair of the world" or "tik-kun olam" as it's termed in the Jewish tradition. And she quotes Michael Bennett who argues that "environmental literature and criticism must needs respond not only to the sublime alterity of the 'black hole of a weasel's eye,' but also to 'the just-closed eyes of a child of the ghetto killed by lead-poisoning from ingesting the peeling paint in his/her immediate environment.'"[6]

Jan Zwicky asks a key question in "The Green Imagination" issue of *The Malahat Review*: "The crux of the issue here is: how can literary writers become advocates for the natural world without being drawn into the propagandizing which undermines the artistic credibility which earns them attention?"[7]

In other words, we know that great nature writing involves all the elements that make a story captivating and gripping: character, plot, narration, conflict and resolution. Because we can never know the mind of the "other" non-human character in our stories, the main character is us, our human interactions, our understanding, our questions, our ideas, our thoughts and our resolution to our own contradictory puzzles.

Thus, what we are really discussing, in some sense, besides the politics of representation and the role of writers who are doing that representation, is the role that writing has in shaping our cultural, social and political ideas about the interaction of humans with the non-human world. So, given both the historical and present, the conflicting ideas in cultural discourse about how we can or should regard the non-human world, given the changing nature of stylistic tropes and paradigms, how do we as writers proceed? I would say, carefully and thoughtfully, with attention to research, to knowledge, to change—with some crucial awareness of what is at stake, and with great attention to our own cultural, personal, social-scientific assumptions and biases.

None of this is prescriptive. There is no right way to do "nature writing." But there is intelligent writing that takes the ethical and cultural implications of these kind of human impacts and human/non-human relationships into account. There is research. There is information. There is story. And good writing that combines these elements is writing that is knowledgeable, thoughtful and questioning.

Or as Jan Zwicky says:

> Conservatism, as in leaving the wilderness alone, isn't the only way of protecting nature. Human language is key, not only because it's an extension of the natural world, but because, when used at full stretch, it has the capacity to change social perception and, potentially, behaviour.[8]

Such writing is what I look for, what I track down, hunt for, devour… it is indeed, following up this metaphor, food for my mind and spirit, and solace for the fear and anguish I feel in today's world. I find it many places and in many kinds of writing. Mostly it is contemporary but not always. Whether it has any answers or solutions for me or for the world is not the point. I want to be part of a writing community that isn't afraid to ask the necessary, the important, the crucial and the world-changing questions.

Solitude

After a summer full of visitors, I am back into my fall writing-walking solitude. On the hill just to the north of our farm are the remains of a man named Bill Haley. Bill lived alone, albeit with a herd of goats for company. There are lots of stories in our area of people who left family or friends to live alone. It was the pioneer era, and so many people were naturally separated from everything they had known. I have often wondered how that was bearable. What was it like, in the days when a letter took three to six months to arrive, for people to be separated for the rest of their lives from everything they had once known as familiar and dear? I know it still goes on but at least now, mail is almost instantaneous. What happens to people who voluntarily choose solitude—monks in caves, hermits in their wilderness?

In winter, I often spend several days alone before the weather clears and I can go to town. I look forward to this and then I dread it. The nights come down hard, dark, quiet. No cars on the road. Electronic people on Facebook are not real people but only memories of people. Emails come in and disappear but they aren't people either, only messages from somewhere suddenly far away. The house is silent. The darkness outside is silent.

The cat curls on my lap. The dog sighs and stretches out on the rug. The future vision that some science fiction writers have had, of people living in self-contained rooms, surrounded by electronic toys, is so impossible that it is almost funny. People simply wouldn't survive like that, or at least, not for long. I write and read and edit and then I drive to town to go for coffee, to the library and come home again to silence.

Harvest

At night I sit outside by the fire, reluctant to go in the house, already claustrophobic at the idea of winter, already missing the dying garden. The light fades, darkness creeps closer. A black-blue storm builds over Midge Creek, behind MacGregor Mountain. I am lonely and glad to be alone, lonely for a place I don't know, lonely for people I haven't met, for children born and unborn, for books unread and books unwritten. The rain and lightning comes.

In the morning, the light is clear on the mountains across the lake. Every tree stands out. The wind is beating the lake hard into silver, the first few turned leaves gold. I wander through tall grass blazing full of points of light. There are two bear scats on the road. The grapes shine a deep maroon, peaches and apples, scarlet and pink, eggplants, a radiant black-purple. All morning, alone, I harvest fruit, put it in buckets and carry it inside for winter.

Staying Put

LATE OCTOBER, ALMOST NOVEMBER, AGAIN. Today it rained for the first time in almost a month and the dogs and I went walking on the mountain through the soft rain. I thought of that wonderful Patrick Lane line, "the bare plum of winter rain."[1] I walked on the steep logging road my father made thirty years ago. I could taste this rain on my skin. The dry ground sucked it down. The tops of some cedars were brown, the moss crackled underfoot, the leaves on the brush were curled and dry. The mountains were a dim grey-blue, the lake surface flat.

When I came back from walking, I sat on the deck for a long time. I had a cup of tea and a book but they sat beside me. I wasn't meditating. I was sitting. There's almost no noise this time of year. Very occasionally, a car or a group of cars went by on the highway, a long wet swish roar. I sat until time dissolved and went away. I sat until I was cold and hungry and then I went inside.

The birds have mostly gone south. The gophers are in hibernation. In succession I said goodbye to the swallows, the ospreys, the geese. In truth I was ready at the end of August for summer to end, for the long push—to get the garden in, and then harvest it and then take it down—to be done.

In the second week of September, finally, silence: WWOOFers, grandkids, visitors gone home, boats gone off the lake, traffic off the road. My freezer is full, my wood is piled. But there is still so much more to do. There are still squash piled on the ground, plants to move into the greenhouse, last-minute chores before winter.

And then time to write again. Maybe.

From the outside, my life looks idyllic and romantic. Sometimes it even feels that way. But I am also often stretched thin by the lines I walk and the demands made in my separate worlds. Writing and farming do not go well together. Neither one pays much, for one thing.

The truly idyllic life would be the one the British aristocracy invented for themselves—an income, plus a cook, a gardener, and perhaps a farmhand to touch his cap and take orders. Aaah yes, and I could stump about, surrounded by dogs, saying, "Put that there and dig that and kill that," and then go off to my oak-lined study to write.

That's the life in the occasional films I like to watch, what my daughter calls "stately mansion full frontal porn," British BBC productions, history, big houses, lots of horses and vast sweeps of lawn. The plot is secondary. I have a suspicion that people expect my life to be like that and don't want to hear that it isn't.

—∞—

All my life, I have shuffled and shuttled back and forth from the farm to the city and back. There was no way for me to make a living as a writer in the country, no way to get an education, no way to be part of scholarly community. I needed what I needed: school, people, training, a job. But I wasn't good at cities or people. I was never sure what to say or how to behave. I kept moving home to the farm and leaving again. This is a hard way to live, especially with children.

For a ten-year period, in my forties, after the kids had all left home, I lived at the farm and wrote books and lived on $5,000 to $6,000 a year. Even on a farm, not paying rent and with access to lots of free food, this wasn't enough money. My main expense was my car. I had to buy it tires and fan belts and oil changes. Visiting my mother, who lived across the yard, was my main consolation. I could get warm in her house and she was always good for coffee and cookies.

I got a lot of writing done in this period. This was before email and internet so I didn't have that to distract me. My television was a small black and white portable— I could vaguely make out what was going on and that was good enough. I usually watched it at night just long enough to get sleepy.

And once a week, from September to April, I got in my precious car and drove two hours to the college in the next town and taught for three hours in the evening and then drove home again.

I wasn't good at being poor, but I was afraid and awkward and unsure of what else to do. I kept writing books and publishing books, and I made no money at all from these books. They went out and disappeared into the market somewhere.

I would put myself to sleep dreaming of suicide and loneliness, dreaming of walking in the high mountains in snow and moonlight, dreaming of solitary hideouts in deep woods or on the edges of solitary lakes.

And throughout this time in my life, this difficult time, my family, my lovely quarreling difficult family, was disintegrating around me, all of us feuding, mostly over money, my father wanting to sell the farm, my brother working furiously as a logger and not seeing anyone else's needs, my children moving away, my mother growing increasingly anxious and angry and depressed, the first threads of dementia beginning to affect her.

And finally, to save my own life, I moved away again, went back to university, yet again, got another job teaching, and kept writing.

There is this apocryphal saying by a Crow elder that gets quoted in various ecological writings and whether it was actually said or not doesn't seem to matter.

> You know, I think if people stay somewhere long enough—even white people—the spirits will begin to speak to them. It's the power of the spirits coming from the land. The spirits and the old powers aren't lost, they just need people to be around long enough and the spirits will begin to influence them.[2]

I don't know about the spirits of the land because I don't have that kind of aboriginal sense. But I do know that living here is essential for me. Without this place, I would be an empty bag of skin walking around. Without this place, I would have no being at all. I belong to it in a spiritual sense and it belongs to me in a financial and temporal sense. Very odd relationship, that one.

But staying in place, of course, runs counter to the North American dream of independence, making lots of money, self-creation, hustling, moving, re-creation of self and family and onwards. The dream encourages people to wander on and re-create who and what they are over and over. But living here has taught me to love every blade of grass, every insect, every tree, to wander around, to live outside time, to be irritated and suspicious of strange people, to farm, to grow food, to listen to every noise, to live intensely with animals, to listen to swallows and crickets and frogs and ospreys and say hello and goodbye to them at the right times.

None of this is worth any money; none is translatable into values recognized by mainstream society. So when I stay home, as I do now, I get a lot of unquantifiable benefits. I get to go outside and be eccentric and wave at ravens. I get to be poor again and have time to write and dream. I get to live in a world of flowers and plants and gardens and neighbours who come by if I need them. I get to think about my grandchildren living here without me. I get to plant trees and wonder what they will look like in a hundred years. I get to dream.

All good and all romantic. And here's the rub. It isn't romantic or idyllic. It isn't stately mansions. It's dirt and work and food. It's ordinary. And it has a price, just not the one people usually imagine.

In the time when Friedrich Engels coined the phrase, "the idiocy of rural life," poor people who lived rurally were part of a class system. The romance and idyllic ideas of rural life came from poets and the upper class.

These days, when I read discussions on the internet about the necessity of reinventing food and agriculture in the days of declining oil supply, I am amazed at how little people seem to understand about the nature of small, mixed, sustainable farming. It's surprising given that many people have grandparents who were farmers. And given that both Canada and the United States were pioneered and settled by people who had to be independent, self-sufficient and skilled in multiple ways, how can we have forgotten this so quickly?

Any conversation about small farming runs up immediately against a soup of contradictions: it is idyllic, romantic—no, it is backbreaking work—lonely, dirty, smelly, germy, a long fight against the forces of nature—no, it is being one with the land, close to the land, learning from the land.

And of course, as is usual with clichés, all of these contain a small kernel of truth and not much more than that. And in fact, within all these small kernels of truth is the reality that not much has changed in the rural parts of North America and until there is some kind of real apocalyptic crisis, it isn't likely to.

In both the US and Canada over the last thirty years, the rural population has mostly fled to the cities. In my community, and in many others, they have been replaced with summer people or tourists, there to have fun, not to live and work. The services, the amenities, the educational facilities, and most of all, the jobs and money, are in cities. It is still impossibly difficult to make a living as a small farmer, although a small farmer can live and eat well and subsist. So for people who choose to stay, who choose land, who choose the place they love, who choose actual rurality, they choose it over career, education and their chance to advance up any kind of economic ladder. Nor will they be part of a rural community. Instead, they will likely be looked at strangely by their new urban neighbours.

Barbara Ehrenreich in her book *Bright Sided*, about the negative side of positive thinking, writes about the amount of leisure time that people living in medieval villages actually had.[3] Farmers, except at specific times of the year such as planting and harvest, could work three or four days a week and still have time for festivals and celebrations. Village life was often marked by holidays, fetes, celebrations, religious rituals and community events far more than it is today. In fact, rural life often tended to be fairly celebratory. What made it difficult wasn't the nature of rural life but the nature of the class system that prevented people from getting an education or better health care, or being literate or mobile.

A healthy functioning rural community that has access to good education, where people are socially and communally minded, would be a good place to live, a good place to raise a family, a good place in which to learn and understand the intricate web of economic, cultural and ecological relationships that connect humans to the places where they live. Industrialization, industrial agriculture, urbanization and suburban ecological deserts have almost destroyed this life, but not quite. Many people miss it and they want it back. They may not even know what it is they miss. But the impulse to form community and to love where one lives is a deep and basic human instinct.

If any of the multiple apocalyptic catastrophes being predicted come about, then it is indeed possible that small farming and rural community, a return to true "peasantry"—meaning being from a region or a rural district—may again arise, may indeed be the saving of people. But that is all in the future.

For now, I and many other rural people survive in a fragmented and class-driven rural society where, unless someone already has money and education, opportunities for education, health care, a decent job and the ability to make any money as a farmer are still very limited. The price is paid in travel, in time, in being split between here and there, urban and rural, and in watching our children and grandchildren go away and be sucked into the busy-ness and madness of cities, of progress, of "careers," all with a price to pay as well.

⸺

I recently spent a month in a writing retreat, in a small town in another province. My reasons for doing this were sound. I would have time to write, walk, read, do research. What a good idea. The small town was friendly, the people were warm, the house was comfortable. I had a house-sitter. The farm was being well cared for. I was miserable. I wanted to go home. My animal self knew what it needed and would not be comforted. I was like a dog in a kennel—fed, housed, unable to relax. I did a lot of writing. I worked hard. When I wasn't writing, I paced the floor and went for walks, and when the retreat was done I fled for home.

White Peaches

I spent two days at the Banff Arts Centre, at a creative non-fiction conference. I dressed in my good clothes. I hung out with writer friends, ate and drank too much, and talked books, writing, publishing and writer gossip. On Sunday afternoon, I drove home through the Rockies. The side of the road was lined with deer and elk, grazing on the new spring grass. Four hours later, with one stop for a box of mixed doughnuts for the freezer, I was home.

The farm was peaceful, the sun was shining, but no one was about. I unpacked, drank tea on the porch and phoned my kids to let them know I was home and safe. About 4 p.m., my brother and our friend Jon arrived to plant a white peach tree. After the tree was planted, Jon and I drank wine on the porch.

We toasted the tree, now named The Malcolm, in honour of Jon's partner, who has cancer. We talked about whether Jon and Malcolm should get married. Brad, my nephew, and a couple of his friends arrived. I brought some beer and smokies up from the basement and they built a fire. We all drank beer and burned hot dogs. The talk was of weather, gardens, animals and local gossip. Jon had brought some fireworks, including an enormous Roman candle. He propped it on a rock, lit it, and it promptly fell over and shot flaming balls of green and purple and red over the field below us.

After everyone left, I lay awake thinking about the many and endless contradictions in my life, and listened to rain pounding on the roof.

Patterns

October whizzed by in patterns of dark and light, rain and sun, cold and warmth. Today, the beauty of this place is like a shout, like a hurrah, it is so bright and astonishing. Dark royal-blue lake, gold leaves, smoky blue air, people coming and going, pigs being carted off to the butcher.

Whenever I get seriously into writing, the farm, the house, my life seems to disintegrate around me. I wake up to dog puke on the rug and piles of paper fluttering to the floor and dust and dying plants even though I have only been "gone" a couple of hours and not really gone at all—just my mind and spirit and perhaps some form of energy that keeps it functioning. I've noticed this before, how much the farm is like a live creature, a creature of spirit and energy, and how when my dad got old and discouraged, some feeling that used to animate the farm and connect it together faded and almost disappeared. The more people and energy there are about the place, the more alive it becomes. So then I think I can either be a writer or I can be a farmer but stubbornly and idiotically, I persist in both. And stubbornly and idiotically, it does work, most of the time. Just far more slowly than I would like.

And of course, the world creeps on, getting stranger and stranger.

What an odd time it is—but it has been odd for so long it has become normal, this split between realities, between this time and future time, between what mass society appears to believe and what prognosticators tell us is going to happen.

And me, I live on the farm and wander about and wonder, pretty much every day, why the beauty of this world and the abundance and wonder and amazing diversity of animals and plants and clouds and weather and gold and blue October mountains isn't enough for people. The beauty of the world— the mystery of it all, and here briefly, us, alive and noticing, paying attention.

Endnotes

FRESH TRACKS

1. Alexandra Horowitz, *Inside of a Dog* (New York: Scribner, 2009).
2. Rebecca Solnit, *Wanderlust: A History of Walking* (New York: Penguin Books, 2000).

INTERBEING/ANIMALIA

1. Edward O. Wilson, *Biophilia* (Cambridge: Harvard, 1984).
2. Jakob von Uexküll, "A Stroll Through the Worlds of Animals and Men" *Instinctive Behavior: The Development of a Modern Concept* (New York: International Universities Press, 1934).
3. Von Uexküll, "A Stroll."
4. Ibid.
5. Suzanne Goldenberg, "BBC 'Bear Man' Documentary Explodes Honey Myth," *The Guardian*, (London), Oct 27, 2009.
6. Yi-fu Tuan, *Landscapes of Fear* (New York: Pantheon, 1970).

7. Craig Childs, *The Animal Dialogues: Uncommon Encounters in the Wild* (New York: Little, Brown and Company, 2007), 13.

8. Barbara Noske, *Beyond Boundaries: Humans and Animals* (Montreal: Black Rose Books, 1997), 170.

9. Jeremy Narby, *Intelligence in Nature* (New York: Penguin Group, 2005).

10. Ibid.

11. Nigel Rothfels, introduction to *Representing Animals* (Bloomington: Indiana University Press, 2002), xi.

WRITING LAND INTO LANGUAGE

1. Eudora Welty, "Place in Fiction," in *Eudora Welty: Stories, Essays and Memoir* (New York: Literary Classics of the United States, 1998).

2. William Fiennes, *The Snow Geese* (New York: Random House, 2002), 122.

3. Fiennes, *Snow Geese*, 122.

4. Jeannette Armstrong, "Keepers of the Earth," Jeannette Armstrong, 1995. Accessed at http://www.sacredland.org/PDFs/KeepersoftheEarth.pdf

5. Glenn Albrecht, "Solastalgia: The Origins and Definition," Healthearth Blog (Jan 12, 2008). Accessed at http://healthearth.blogspot.com/search/label/Solastalgia)

6. Albrecht, "My Take on Occupy Wall Street," Healthearth blog (Nov 5, 2011).

7. Jeannette Armstrong, "I Stand With You Against the Disorder," *Yes Magazine* (Nov 08, 2005). Accessed at http://www.yesmagazine. org/issues/spiritual-uprising/i-stand-with-you-against-the-disorder2005)

THE PLACE LOOKS BACK

1. Jay Appleton, *The Experience of Landscape* (London: John Wiley), 69.
2. Appleton, *Experience*, 73.
3. John Berger, *Ways of Seeing* (London: Penguin Books, 1972), 148.
4. Gary Snyder, *The Practice of the Wild* (New York: North Point Press, 1990), 39.
5. Snyder, *Practice*.
6. Ibid.
7. John Berger, *Seeing*.

WRITING AFTER NATURE

1. Kate Rigby, "Writing After Nature," *Australian Humanities Review*, no. 39 – 40 (2006). Accessed at http://www.australianhumanitiesreview.org/archive/issue-September-2006/rigby.html
2. Ibid.
3. Luanne Armstrong and Zoe Landale, *Slice me some truth* (Hamilton, Wolsak & Wynn, 2011), 16.
4. Gary Snyder, *Back on the Fire* (Berkeley, CA: Counterpoint Press, 2007), 24.

5. Cees Nooteboom, *Writers and Company*, CBC
    Radio, March 21, 2010. http://www.cbc.ca/
    writersandcompany/episode/2010/03/21/sunday-
    21-march-2010-and-wednesday-24-march-
    2010-cees-nooteboom/

6. Kate Rigby, "Dancing with Disaster," *Australian
    Humanities Review*, no. 46 (2009). Accessed at
    http://www.australianhumanitiesreview.org/
    archive/issue-May-2009/rigby.html

7. Jan Zwicky, "The Details: An Interview with
    Jan Zwicky," *The Malahat Review* Green
    Imagination Issue, no. 165 (2008).

8. Zwicky, *Malahat*.

STAYING PUT

1. Patrick Lane, *The Bare Plum of Winter Rain*
    (Madeira Park, BC: Harbour Publishing, 2010).

2. Crow elder, quoted in *The Practice of the Wild*,
    Gary Snyder.

3. Barbara Ehrenreich, *Bright Sided* (New York:
    Metropolitan Books, 2009).

Thank You

THANKS ARE DUE TO MANY people who assisted with this book: my brother Bill and sister, Robin, my writer's group—Linda Breault, Kelly Ryckman, Ilana Cameron and Tanna Patterson—plus for reading and comments Holley Rubinsky and Kuya Minogue. K. Linda Kivi gave much help, conversation and editing support. Thanks to my wonderful editor, Jane Silcott, and to Vici Johnstone at Caitlin Press for her fine work and her belief in writers and writing.

Luanne Armstrong

LUANNE ARMSTRONG LIVES IN THE small community of
Boswell, BC, where she farms land with her siblings. She
has worked as a feminist researcher, a freelance journalist
and a writing instructor. She teaches Creative Writing at
the College of the Rockies.

More Non-Fiction from Caitlin Press

TSE-LOH-NE: THE PEOPLE AT THE END OF THE ROCKS
Keith Billington

Billington was invited on a traditional Sekani trek along the Aatse Davie Trail using pack dogs, traversing 460 kilometres in some of BC's roughest terrain.   .

9781894759885 / $22.95 / 60 B&W photos

THE EARTH REMEMBERS EVERYTHING
Adrienne Fitzpatrick

The Earth Remembers Everything is a masterful blend of history, travel and poetic narrative, tracing the author's journeys to some of the world's most difficult destinations.

9781894759908 / $18.95

WOMEN OF BRAVE METTLE
Diana French

In this much-anticipated second volume in the Extra-ordinary Women series, Diana French follows up with more stories of the women of the Cariboo Chilcotin.

9781894759861 / $26.95 / 60 B&W photos